# Reviews of Environmental Contamination and Toxicology

## VOLUME 170

**Springer**
*New York*
*Berlin*
*Heidelberg*
*Barcelona*
*Hong Kong*
*London*
*Milan*
*Paris*
*Singapore*
*Tokyo*

# Reviews of Environmental Contamination and Toxicology

Continuation of Residue Reviews

Editor
George W. Ware

VOLUME 170

Springer

# Coordinating Board of Editors

Springer-Verlag
*New York*: 175 Fifth Avenue, New York, NY 10010, USA
*Heidelberg*: Postfach 10 52 80, 69042 Heidelberg, Germany

Printed in the United States of America.

ISSN 0179-5953

Printed on acid-free paper.

ISBN 978-1-4419-2925-9

Springer-Verlag New York Berlin Heidelberg
*A member of BertelsmannSpringer Science + Business Media GmbH*

# Foreword

International concern in scientific, industrial, and governmental communities over traces of xenobiotics in foods and in both abiotic and biotic environments has justified the present triumvirate of specialized publications in this field: comprehensive reviews, rapidly published research papers and progress reports, and archival documentations. These three international publications are integrated and scheduled to provide the coherency essential for nonduplicative and current progress in a field as dynamic and complex as environmental contamination and toxicology. This series is reserved exclusively for the diversified literature on "toxic" chemicals in our food, our feeds, our homes, recreational and working surroundings, our domestic animals, our wildlife and ourselves. Tremendous efforts worldwide have been mobilized to evaluate the nature, presence, magnitude, fate, and toxicology of the chemicals loosed upon the earth. Among the sequelae of this broad new emphasis is an undeniable need for an articulated set of authoritative publications, where one can find the latest important world literature produced by these emerging areas of science together with documentation of pertinent ancillary legislation.

Research directors and legislative or administrative advisers do not have the time to scan the escalating number of technical publications that may contain articles important to current responsibility. Rather, these individuals need the background provided by detailed reviews and the assurance that the latest information is made available to them, all with minimal literature searching. Similarly, the scientist assigned or attracted to a new problem is required to glean all literature pertinent to the task, to publish new developments or important new experimental details quickly, to inform others of findings that might alter their own efforts, and eventually to publish all his/her supporting data and conclusions for archival purposes.

In the fields of environmental contamination and toxicology, the sum of these concerns and responsibilities is decisively addressed by the uniform, encompassing, and timely publication format of the Springer-Verlag (Heidelberg and New York) triumvirate:

*Reviews of Environmental Contamination and Toxicology* [Vol. 1 through 97 (1962–1986) as Residue Reviews] for detailed review articles concerned with any aspects of chemical contaminants, including pesticides, in the total environment with toxicological considerations and consequences.

*Bulletin of Environmental Contamination and Toxicology* (Vol. 1 in 1966) for rapid publication of short reports of significant advances and discoveries in the fields of air, soil, water, and food contamination and pollution as well as methodology and other disciplines concerned with the introduction, presence, and effects of toxicants in the total environment.

*Archives of Environmental Contamination and Toxicology* (Vol.1 in 1973) for important complete articles emphasizing and describing original experimental or theoretical research work pertaining to the scientific aspects of chemical contaminants in the environment.

Manuscripts for *Reviews* and the *Archives* are in identical formats and are peer reviewed by scientists in the field for adequacy and value; manuscripts for the *Bulletin* are also reviewed, but are published by photo-offset from camera-ready copy to provide the latest results with minimum delay. The individual editors of these three publications comprise the joint Coordinating Board of Editors with referral within the Board of manuscripts submitted to one publication but deemed by major emphasis or length more suitable for one of the others.

Coordinating Board of Editors

# Preface

Thanks to our news media, today's lay person may be familiar with such environmental topics as ozone depletion, global warming, greenhouse effect, nuclear and toxic waste disposal, massive marine oil spills, acid rain resulting from atmospheric $SO_2$ and $NO_x$, contamination of the marine commons, deforestation, radioactive leaks from nuclear power generators, free chlorine and CFC (chlorofluorocarbon) effects on the ozone layer, mad cow disease, pesticide residues in foods, green chemistry or green technology, volatile organic compounds (VOCs), hormone- or endocrine-disrupting chemicals, declining sperm counts, and immune system suppression by pesticides, just to cite a few. Some of the more current, and perhaps less familiar, additions include *xenobiotic transport, solute transport, Tiers 1 and 2, USEPA to cabinet status, and zero-discharge*. These are only the most prevalent topics of national interest. In more localized settings, residents are faced with leaking underground fuel tanks, movement of nitrates and industrial solvents into groundwater, air pollution and "stay-indoors" alerts in our major cities, radon seepage into homes, poor indoor air quality, chemical spills from overturned railroad tank cars, suspected health effects from living near high-voltage transmission lines, and food contamination by "flesh-eating" bacteria and other fungal or bacterial toxins.

It should then come as no surprise that the '90s generation is the first of mankind to have become afflicted with *chemophobia*, the pervasive and acute fear of chemicals.

There is abundant evidence, however, that virtually all organic chemicals are degraded or dissipated in our not-so-fragile environment, despite efforts by environmental ethicists and the media to persuade us otherwise. However, for most scientists involved in environmental contaminant reduction, there is indeed room for improvement in all spheres.

Environmentalism is the newest global political force, resulting in the emergence of multi-national consortia to control pollution and the evolution of the environmental ethic. Will the new politics of the 21st century be a consortium of technologists and environmentalists or a progressive confrontation? These matters are of genuine concern to governmental agencies and legislative bodies around the world, for many serious chemical incidents have resulted from accidents and improper use.

For those who make the decisions about how our planet is managed, there is an ongoing need for continual surveillance and intelligent controls to avoid endangering the environment, the public health, and wildlife. Ensuring safety-in-use of the many chemicals involved in our highly industrialized culture is a dynamic challenge, for the old, established materials are continually being displaced by newly developed molecules more acceptable to federal and state regulatory agencies, public health officials, and environmentalists.

Adequate safety-in-use evaluations of all chemicals persistent in our air, foodstuffs, and drinking water are not simple matters, and they incorporate the judgments of many individuals highly trained in a variety of complex biological, chemical, food technological, medical, pharmacological, and toxicological disciplines.

*Reviews of Environmental Contamination and Toxicology* continues to serve as an integrating factor both in focusing attention on those matters requiring further study and in collating for variously trained readers current knowledge in specific important areas involved with chemical contaminants in the total environment. Previous volumes of *Reviews* illustrate these objectives.

Because manuscripts are published in the order in which they are received in final form, it may seem that some important aspects of analytical chemistry, bioaccumulation, biochemistry, human and animal medicine, legislation, pharmacology, physiology, regulation, and toxicology have been neglected at times. However, these apparent omissions are recognized, and pertinent manuscripts are in preparation. The field is so very large and the interests in it are so varied that the Editor and the Editorial Board earnestly solicit authors and suggestions of underrepresented topics to make this international book series yet more useful and worthwhile.

*Reviews of Environmental Contamination and Toxicology* attempts to provide concise, critical reviews of timely advances, philosophy, and significant areas of accomplished or needed endeavor in the total field of xenobiotics in any segment of the environment, as well as toxicological implications. These reviews can be either general or specific, but properly they may lie in the domains of analytical chemistry and its methodology, biochemistry, human and animal medicine, legislation, pharmacology, physiology, regulation, and toxicology. Certain affairs in food technology concerned specifically with pesticide and other food-additive problems are also appropriate subjects.

Justification for the preparation of any review for this book series is that it deals with some aspect of the many real problems arising from the presence of any foreign chemical in our surroundings. Thus, manuscripts may encompass case studies from any country. Added plant or animal pest-control chemicals or their metabolites that may persist into food and animal feeds are within this scope. Food additives (substances deliberately added to foods for flavor, odor, appearance, and preservation, as well as those inadvertently added during manufacture, packing, distribution, and storage) are also considered suitable review material. Additionally, chemical contamination in any manner of air, water, soil, or plant or animal life is within these objectives and their purview.

Normally, manuscripts are contributed by invitation, but suggested topics are welcome. Preliminary communication with the Editor is recommended before volunteered review manuscripts are submitted.

Tucson, Arizona                                                      G.W.W.

# Table of Contents

Rev Environ Contam Toxicol 170:1–11

# Fire and Ecotoxicological Aspects of Polyurethane Rigid Foam

### Friedrich-Wilhelm Wittbecker and Manfred Giersig

## Contents

## I. Introduction

Polyurethane (PUR) rigid foam is used as the insulating layer for facades and roofs, especially in cold storage facilities. The material is a high molecular cross-linked thermoset plastic that shows chemical resistance under normal conditions in handling and usage. Owing to its stability, rigid PUR foam is virtually not bioavailable and hence not dangerous according to relevant European Union (EU) regulations and directives.

Normally, PUR rigid foams have facings that are applied without any additional adhesives to produce insulating boards and sandwich elements that are load bearing and immediately ready for use. PUR rigid foams contain closed cells and, because of their physical and chemical properties (Heilig et al. 1992), they are preferably applied in areas with high thermal insulating requirements.

Ecotoxicological aspects such as the toxic potency of fire effluents, the generation of dioxins, and recycling are rarely considered at the design stage of a building. Requirements in prescriptive codes are not clearly defined, and they are not incorporated in reaction to fire classifications, mainly because the primary fire safety objective is to ensure the escape and rescue of occupants. How-

Communicated by George W. Ware.

F.-W. Wittbecker (✉)
Fire Technology Department, Bayer AG, 51368 Leverkusen, Germany.

M. Giersig
Department of Environmental and Product Safety, Bayer AG, 51368 Leverkusen, Germany.

ever, ecotoxicological aspects are becoming more important, and the following ecotoxicological aspects of rigid PUR foam are addressed in this contribution:

Fire performance
Acute toxic effects of fire effluent on humans and the environment
Generation of dioxins in a fire
Usage of partly halogenated hydrocarbons (HCFCs) and
  nonhalogenated blowing agents
Legal situation
Influence of the blowing agent on the fire performance
Conservation of resources and recycling

## II. Fire Safety
### A. General

A fire safety assessment begins by identifying the fire safety objective(s) and acceptable levels of safety (ISO CD 1579: 1999 guidance on intermediate scale fire testing of plastics). The primary fire safety objective is to ensure evacuation of the compartment before untenable conditions are created. The evacuation time includes the time required for the occupants to reach a safe location. Tenability is assessed on the basis of fire effects on the occupants, including both direct effects such as heat, toxic gases, and oxygen deprivation, and indirect effects such as reduced visibility due to smoke obscuration. Additional fire safety objectives, which are designed to prevent serious injury for fire fighters, are also considered.

A number of fire parameters influence the development of a fire (Table 1). Fire parameters between the preflashover and the postflashover conditions are distinctly different. There are three main stages of fire development within an enclosure: initiation, developing fire, and the developed fire. In the preflashover phase, the reaction to fire characteristics of products are important, whereas in the postflashover phase resistance to fire parameters of complete assemblies apply.

### B. Fire Performance

A number of phenomena are known to influence the reaction to fire characteristics of products in the preflashover phase. The response of products to develop-

Table 1. Fire parameters related to classification and testing.

| Fire situation | Stage | Parameters |
|---|---|---|
| Preflashover | Initiation | Ignitability |
| | Developing fire | Fire growth (ignitability, flame spread, heat release) |
| Postflashover | Developed fire | Resistance to fire (load-bearing, integrity, insulation) |

ing fire conditions is evaluated using tests for ignitabilty, flame spread, and rate of heat release. The resistance of structures affected by fire is evaluated using tests based on the ISO 834 time–temperature curve. Performance is classified in terms of fire resistance times for retention of load-bearing capacity, integrity, and separating function.

Ease of ignition is fundamental to all fire situations and depends on the ignition sources present. Noncombustibility is understood to represent very limited contribution to a fire from substantially inert materials such as metals and inorganic substances. The typical ignition source is small, such as candles, matches, or hot electrical wires. Flame-retarded PUR rigid foams are resistant against these primary ignition sources. The ease of ignition under additional radiation is also important to items adjacent to a burning object, as well as in cases where the fire originates outside the enclosure (e.g., fire spreads down a corridor from an adjacent room fire).

The rate of flame spread is important in all scenarios, especially when the spread is wind aided (i.e., vertically up walls and facades and horizontally under ceilings where propagation distances can be extensive). Wind-opposed flame spread is relevant in cases in which there is potential for large spreading distances, as in large rooms and corridors. Generally, self-sustained smoldering may be propagated inside structural elements, but PUR rigid foams do not exhibit such performance.

The rate of heat release is particularly important in small enclosures and where the heat released is accumulated within a building. Heat release is an additional parameter for predicting fire growth. The net calorific value of PUR rigid foam is about 6.7 kWh/kg.

Smoke becomes important if it collects in a building, especially if it reduces visibility in following escape routes. Flaming debris may pose an additional risk of a discontinuous flame spread and rapid propagation of the burning area. PUR rigid foams, however, do not produce flaming droplets.

Fire resistance is defined as the load-bearing capacity of columns, beams, or walls or the ability of a boundary component of a compartment to hold back the spread of a fully developed fire to an adjacent compartment. There is no indication in fire statistics that PUR rigid foam is the first ignited item, nor does it spread the fire to adjacent areas. If flashover occurs in the fire compartment, the foam contributes to heat release and the yield of effluents.

## C. Normative Fire Testing and Classification

Fire tests vary considerably and focus on different simulated risk situations such as walls and roofs. Depending on the test procedure and end-use conditions for composite elements, this may lead to different classifications, although the material composition is chemically identical.

A product can be used in buildings and structures if it complies with the regulatory requirements. Standardized test methods are used to assess the performance of construction products in terms of ignitability, surface spread of flame, and formation of flaming drips by way of small-scale reaction-to-fire test methods, which differ from country to country. Depending on the test method, the relevance of the fire parameters being assessed is different. The tests are performed either on the material or on a small- or intermediate size composite.

Generally, the reaction to fire performance depends on a variety of parameters, such as shape and distribution of the material, ventilation, and end-use conditions. Material-related measurements in small-scale tests might be grossly misleading in terms of a risk assessment. Reaction to fire tests and assessments should therefore take into account intended end-use conditions as much as possible. Some tests measure only one of the characteristics while others measure several at the same time. Table 2 gives an overview of national fire tests; about 40 different reactions to fire tests are used for building products today.

Several classification systems require different levels of performance for composite elements: one for the foam itself and a second one for the end-use product. PUR rigid foams are normally applied in cavities or behind noncombustible protective surface coverings.

The minimum contribution to fire classification for building products is usually determined by combustibility tests such as ISO 1182 (furnace test for homogeneous materials) and ISO 1716 (bomb calorimeter). In some countries all materials for construction must have a minimum performance. In contrast, other countries such as France, Spain, the UK, and Benelux do not require minimum performance levels so long as the end-use building elements meet the building fire regulation requirements (Wittbecker et al. 1999).

Almost all tests focus on the vertical sample orientation. An international standard, Comité Européen de Normalisation or International Organization for Standardization (CEN or ISO), for testing reaction of roofs to a fire inside a building is not available. It is recognized that standards for facades and building envelopes are not being developed on an international basis; ISO TC92/SC1/WG7 has a mandate for developing tests for facades and sandwich panels. At the same time insurance companies such as LPC and Factory Mutual are also developing tests. Three reaction-to-fire tests are being developed in CEN for the impact from the outside of a building.

Links between test requirements and the performance characteristics of construction products need to be established for walls, ceilings, and floors (including their surface coverings); materials incorporated within building elements or cavities; externally applied insulation for facades and external walls; and pipe and duct components.

One product may appear under more than one application; this can lead to different levels of performance of a product in a given structure because of the different fire scenarios associated with the intended use. Thus, products need to be assessed according to end-use conditions in different scenarios.

PUR sandwich panels have also been tested extensively in large- and real-scale tests (Wiese and Wittbecker 1995; Sommerfeld et al. 1996; Jagfeld 1988; Cope et al. 1995), and it has been demonstrated that with impinging realistic

Table 2. A comparison of national fire tests in Europe.

| Subject of test | Germany | France | Italy | Nordic countries | UK |
|---|---|---|---|---|---|
| Reaction to fire: noncombustible | DIN 4102–1 furnace test | NF P92–501 | | NT Fire 001 | BS 476; Pt. 11 |
| Calorific value or potential | | | UNI 7557 (1976) | | |
| Ignitability | DIN 4102–1 | NF M3–M5 NF P92–501 | CSE RF 1/75/A and 2/75/A | NT Fire 002 | BS 476; Pt. 5 BS 476; Pt. 13 BS 476; Pt. 12 |
| Spread of flame | DIN 4102–1 | NF P92–501 | CSE RF 3/77 | NT Fire 004 NT Fire 006 NT Fire 007 | BS 476; Pt. 7 BS 476; Pt. 1 |
| Smoke generation | DIN 4102–15 and 16 | | | NT Fire 004 NT Fire 007 | BS 5111; Pt. 1 BS 6401 |
| Fire resistance based on ISO 834 | DIN 4102–2 DIN 4102–3 DIN 4102–7 | Arrete, 10–9–70, Arrete, 05–2–59 as amended | UNI 7678 (1972) | NT Fire 005 | BS 476; Pt. 3 BS 476; Pt. 22 BS 476; Pt. 21, 23, 24 |

primary ignition sources there is no risk of initiating a fire. Applying large secondary ignition sources directly in front of joints causes the core material to decompose, however, without spreading the fire far from the area of its origin. In fully developed fires (flashover), the core insulating material is largely decomposed and contributes to the fire load. As a result of the decomposition process the insulating performance decreases, and the walls and roof release heat from the fire compartment.

The fire performance of PUR rigid foam was scrutinized after the ban on the use of CFC11 as a blowing agent. Various small-scale test methods have been used as well as realistic simulated fire situations (Heilig et al. 1991; Walter and Wittbecker 1993). For the assessment, a distinction has to be made between the combustion behavior of the blowing agent, which affects the production process, and the performance of the end-use applied product. To evaluate the effects of different blowing agents on the fire performance of the product for the purpose of a general risk assessment, systematic comparisons between CFC and pentane blown foams have been made. Possible impacts of blowing agents from noncombustible to easily ignitable should be covered by this investigation. To conclude the investigation: no general change in fire performance was apparent.

## III. Effects of Fire Effluents on Humans and the Environment
### A. Acute Toxic Effects

All organic materials generate fire effluents of complex composition. Their acute inhalation toxicity is mainly determined by carbon monoxide (CO) (Engler et al. 1990). Moreover, smoke contains primarily carbon dioxide ($CO_2$), water vapor, and soot. Furthermore, nitric oxides ($NO_x$) are generated, together with traces of hydrocyanics (HCN) and isocyanates, in addition to a variety of other components typical of a PUR fire. In the case of polyurethanes containing additives based on halogen or phosphorus, corresponding compounds are also generated. According to Kimmerle et al. (1992) and Pfeil (1968), the HCN yield during smoldering or burning of wood, wool, or leather may be greater than from PURs. For a toxicological assessment, however, single smoke components are not relevant because the total effect of effluents must be assessed. Realistically, this can only be done using animal tests. For basically unknown products, correlations have been established between numerically derived predictions based on analytical results and animal tests. Animal tests can therefore be considerably reduced.

The chemical content and the concentration of fire effluents are dependent on decomposition conditions of temperature and ventilation. These effects are dominant with regard to the magnitude of the acute toxicity of the smoke. Comparative tests taking into account different fire stages have confirmed that the toxicity of fire effluents from PUR products is similar to natural products such as wood, cork, and wool (Effenberger 1973; Kimmerle and Prager 1980; Kimmerle et al. 1992; Prager et al. 1994). Because effluents from PUR do not differ greatly from natural materials, a similar effect on the environment is assumed.

Due to their reactivity, isocyanates are rapidly decomposed photochemically or converted chemically (Duff 1985).

Investigations after fires and numerical distribution predictions have further confirmed that the concentration of effluents decreases exponentially with the distance from the fire source. Concentrations decrease at some distance from the fire down to the sub-ppm level. In the case of a fire in a plastics warehouse, for example, a nitrogen dioxide concentration of about 1.2 ppm is predicted at a distance of 100 m from the fire. At a distance of 200 m from the fire area, the concentration decreases to about 0.4 ppm. The calculation was done according to German VDI (Association of German Engineers) Guideline 3783, Part 1, with the following assumptions: medium distribution situation, 15-min fire duration, 900-kg fire load, no heat emission, 1% yield rate, $NO_2$ source rate 10 g/sec.

In a fully developed fire, conditions for the closer environment becomes even less critical; for example, the $NO_2$ concentration decreases at a distance of 100 m from the fire to less than 0.001 ppm (fire duration, 2 hr; consumed material, 42,000 kg; 45 MW heat emission; yield rate, 0.12%; $NO_2$ source rate, 7 g/sec.

## B. Fires and Yield of Dioxins

In practically all fires in which natural products or plastics are involved, a variety of residual products to which the group of dioxins (polyhalogenated dibenzo-dioxins and -furans) belong are generated in addition to the universally known smoke components (Engler et al. 1990).

During a fully developed fire, elevated temperatures are reached that promote the generation of dioxins if chlorine or bromine is available. Natural products and so-called halogen-free products normally also contain halogenated compounds in the form of technically unavoidable traces. In every fire, the yield of dioxins and furans therefore has to be expected on a sub-ppm level. In addition, dioxins and furans are generated in every controlled combustion process and they are present everywhere in the environment (Roßmann 1996). This background level has to be taken into account when assessing any fire disaster. Fire residues normally contain dioxins whose hazard potential cannot be evaluated on a general basis. The group of dioxins and furans contain 210 different chlorinated compounds and a similar number of bromine derivatives.

Analytical data can only be the basis for a risk assessment if the type of dioxins and the concentration found in the residues are known. About 25 dioxins have been identified as toxicologically relevant; special regulatory measures exist in some countries for these types of dioxins.

The complex mode of action of dioxins has not been finally identified. Because statements concerning hazards of exposure to dioxins are emotional, hazard assessments for humans and the environment should be based on rationally derived regulatory limits. For example, in Germany a threshold limit value of 2 µg/kg exists for TCDD, which is the reference substance for dioxins. Products containing 2 µg/kg TCDD or more must be handled like products classified as human carcinogens. Adequate hygienic precautions have to be taken when handling such products. With the help of internationally agreed toxicological equivalence factors, the most important halogenated dioxins and furans can also be assessed by analogy with TCDD. For the sake of completeness, the biological availability of dioxins bonded to soot is very poor (Roßmann 1996).

In PUR fires, quantities of brominated or chlorinated dioxins or furans are not expected to exceed a total equivalent to 2 µg/kg TCDD; this has been confirmed by tests conducted under controlled laboratory conditions at Bayer AG and in an independent laboratory. PUR rigid foams containing halogenated and halogen-free flame retardants were investigated, with CFC11 being used as the blowing agent in the earlier investigations and HCFC in recent tests. The insulating products were made from raw materials supplied by Bayer AG.

There are only a few analytical investigations concerning realistically large fires. One example is the systematic investigations made during a fire in a plastics warehouse in Lengerich, Germany, in 1992. The consumed fire load was estimated to be between 1500 and 2500 t of mixed PVC-containing recycling materials (Troitzsch 1992). A comparison of data from this fire with data obtained under normal environmental conditions did not show any significant additional pollution due to dioxins. At a distance of about 200–250 m from the fire source, the dioxin concentration in the air was on the same level as that which can be measured in areas of big cities (Troitzsch 1992). Dioxin concentrations measured at the Lengerich fire area are also found in normal domestic dwelling fires (Troitzsch 1992). In addition, dioxin concentrations of residues have been evaluated as not critical (based on the TCDD threshold limit value of 2 µg/kg).

It goes without saying that after every fire precautions have to be taken for appropriate disposal of the waste and for demolition or restoration of the building. Requirements and recommendations are available (Roßmann 1997; Verband der Sachversicherer e.V. 1990). In some countries, special regulations and guidelines exist, for instance, the German TRGS 524: "Redevelopment and Work in Contaminated Areas" (Sanierung und Arbeiten in kontaminierten Bereichen).

## IV. HCFCs and Halogen-Free Blowing Agents for the Production of Rigid PUR Foams

An EU Directive (Council Regulation (EC) No 2037/2000 of 29 June 2000 on substances that deplete the ozone layer) regulates the use of HCFC in Europe. This Directive prohibits the use of HCFCs as blowing agents for the production of rigid PURs after 31 December 2003.

Blowing agents have two functions in PUR rigid foam production. They serve as the blowing agent to expand the foam and remain in the closed cells as an insulating gas contributing to the insulating properties. Diffusion of the blowing agents through the cell membrane is very slow. The half-life of HCFC in PUR rigid foams is longer than the expected lifetime of the product. Surface coverings in the form of steel facings act as an additional protection by preventing the release of the blowing and insulating agents. Insulating properties, therefore, change very little during the lifetime of the building.

In the past, almost all the comparative studies concerning the toxicity of effluents from a fire have utilized rigid foams containing CFC11. It would, however, be justifiable to presume similar results for products based on HCFCs and halogen-free blowing agents.

## V. Conservation of Resources and Recycling

Recently published data (Walter and Weigand 1998) show that the excellent insulation properties of rigid PUR foam contribute significantly to environmental protection and the conservation of resources. For instance, the saving in heating energy during use far outweighs the production "costs" and the energy credit for recovery.

A number of methods have been developed for recycling PURs, some of which have been implemented in practice. The recycling options for rigid foam can be grouped under three main categories:

Physical and mechanical recycling: Pulverized material as a basis for the production of pressboards or as adsorbents; the insulating cell gases can be collected and disposed of separately.

Chemical recycling: Glycolysis (chemical breakdown) of polyurethanes into liquid raw materials, which are then reused for the production of new polyurethanes.

Energy recovery.

Furthermore, there is the possibility of thermal treatment processes, i.e., recovering the calorific content by incineration in suitable plants. Large-scale tests have confirmed that halogenated blowing agents and flame retardant are destroyed in the decomposition process and that the concentration of dioxins in the flue gas is comparable with that of common municipal solid waste incineration plants (Vehlow and Mark 1995).

## Summary

The main characteristics of fire effluents from polyurethane (PUR) foams are comparable to those from natural materials like wood, cork, or wool. This similarity has been demonstrated by comparative data from analytical and toxicological studies. It is therefore presumed that effluents of these materials present similar hazards to human beings and the environment.

In almost all fires, dioxins can be found in the smoke and residues. In fires involving PURs, relevant quantities of halogenated dioxins or furans are not to be expected; this has been confirmed by investigations under controlled laboratory conditions.

The insulation properties of rigid PUR foam contribute significantly to environmental protection and the conservation of resources. A number of methods for reusing and recycling PUR rigid foam waste have been developed and realized in practise. The possibilities range from reusing the material itself, generating liquid raw materials, and thermal recycling, even for (H)CFC-containing PUR rigid foams, by cocombustion in suitable plants.

# References

Cope B, Hildebrand C, Levio E, Prager FH (1995) Fire performance of PUR steel sandwich panels. ISOPA, Brussels.

Duff PB (1985) Fate of toluene diisocyanate in air phase II study. In: Proceedings, SPI 29th Annual Technical Marketing Conference, Reno, NV,

Effenberger E (1973) Gutachten: Zur Frage der Verwendbarkeit von Polyurethan (PUR)-Hartschaumstoff sowie zweier Polyisocyanurat—Schaumstoffe (PIR I und PIR II). der Bayer AG, Leverkusen, im Bauwesen.

Engler A, Pieler J, Einbrodt HJ (1990) Gefährdungsabschätzung von Brandgasen und Brandrückständen unter humantoxikologischen Gesichtspunkten. Wissenschaft und Umwelt 191 no. 4 and references cited in the literature.

Heilig G, Prager FH, Walter R, Wiedermann R, Wittbecker F-W (1991) Brandverhalten pentan-getriebener Polyurethan—Hartschaumstoffe/Burning behaviour of pentane blown polyurethane rigid foams. Kunststoffe, German Plastics, no. 9., pp 790–794.

Heilig G, Prager FH, Walter R, Wiedermann R, Wittbecker F-W (1992) Pentangetriebene Polyurethan (PUR)Hartschaumstoffe—Bauphysikalische Eigenschaften. Bauphysik 3: 76–81.

Jagfeld P (1988) Verhalten von Stahl-PUR Sandwichelementen im Naturbrandversuch. VFDB Zeitschrift 1:10–15.

Kimmerle G, Prager FH (1980) The relative toxicity of pyrolysis products. Part II. Polyisocyanate based foam. J Combust Toxicol 54:54–68.

Kimmerle G, Pauluhn J, Prager FH (1992) Rauchgastoxizität von Polyurethan-Verbrennungs-/Verschwelungsprodukten. Kautschuk Gummi Kunststoffe 45(2):141–148.

Pfeil E (1968) Expertise concerning the HCN yield during combustion of Hartmoltopren (in German). Bayer AG, Leverkusen, Germany.

Prager FH, Kimmerle G, Maertins T, Mann M, Pauluhn J (1994) Toxicity of the combustion and decomposition products of polyurethanes. Fire Mater, 18:107–119.

Roßmann G (1996) Sanierung von Brandschäden. Teil I. Gefahrstoffe nach Bränden. vfdB Zeitschrift 4.

Roßmann G (1997) Sanierung von Brandschäden. Teil II. Gefährdungseinschätzung kalter Brandstellen. VFDB Zeitschrift 1.

Sommerfeld C-D, Walter R, Wittbecker F-W (1996) Fire performance of facades and roofs insulated with rigid polyurethane foam—a review of full scale tests. Interflam Proceedings, pp 315–325.

Troitsch J (1992) Bericht: Brand in Firma Micro-Plast, Lengerich, am 4.10.1992, Nr. TR 042993.

Verband der Sachversicherer e.V. (1990) Leitfaden zur Brandschadensanierung. Empfehlung zur Reinigung von Gebäuden nach Bränden, Bundesgesundheitsamt (BGA) Bekanntmachung. Bundesgesundheitsblatt 33.

Vehlow J, Mark FE (1995) Co-combustion of building insulation foams with municipal solid waste. ISOPA, Brussels.

Walter R, Weigand E (1998) Umweltschutz und Ressourcenschonung durch Wärmedämmung am Beispiel von Polyurethanen (PUR)-Hartschaumstoffen. Bauphysik 20:145–150.

Walter R, Wittbecker F-W (1993) Assessment of PUR Sandwich Panels in Fire Case with Special Regard to Pentane Blown Foam. Interscience, London.

Wiese J, Wittbecker F-W (1995) Verhalten einer Polyurethan-Sandwich-Paneelwand bei lokaler Brandbeanspruchung. VdS Magazin Schadenverhütung + Sicherheitstechnik, no 6, pp 29–31.

Wittbecker F-W, Daems D, Werther H-U (1999) Performance of Polyurethane (PUR) building products in fires. ISOPA, Brussels.

Manuscript received March 27, 2000; accepted June 19, 2000.

Rev Environ Contam Toxicol 170:13–74

# Pesticide Acute Toxicity Reference Values for Birds

Pierre Mineau, Alain Baril, Brian T. Collins, Jason Duffe,
Gerhard Joerman, and Robert Luttik

## Contents

## I. Introduction

Avian risk assessment of pesticides depends for the most part on two laboratory-derived measures of lethality. First, the median lethal dose ($LD_{50}$), a statistically derived single oral dose of a compound that will cause 50% mortality of the test population, and second, the median lethal concentration ($LC_{50}$), which similarly derives the concentration of a substance in the diet that is expected to lead to 50% mortality of the test population. Mineau et al. (1994) have argued against the continued use of the $LC_{50}$ endpoint in avian risk assessment of pesticides. The test as currently designed was found to provide unreliable results, in part because of the difficulty of properly determining exposure during the test. The $LC_{50}$ test, conducted on very young birds, is greatly influenced also by the exact age and condition of the test population. Also, the correlation of $LC_{50}$ values among test species is weak, thus casting further doubt on the value of the endpoints and limiting our ability to extrapolate from test species to wild bird spe-

---

Communicated by George W. Ware.

P. Mineau (✉), A. Baril, B.T. Collins, J. Duffe
National Wildlife Research Centre, Canadian Wildlife Service, Environment Canada, Hull, Québec, Canada, K1A OH3.

G. Joerman
Biologische Bundesanstalt für Land-und Forstwirtschaft, Braunschweig, Germany.

R. Luttik
RIVM, 3720 BA Bilthoven, The Netherlands.

cies. Finally, comparison of test results with field evidence suggests that lab-derived $LC_{50}$s are poor predictors of risk. Until the $LC_{50}$ test is redesigned to address these weaknesses, avian risk assessment will depend almost entirely on the results of the $LD_{50}$ test.

Avian risk is difficult to estimate in an absolute sense from laboratory-derived data. Field studies are generally needed to provide "ground truthing" of a risk model and, to date, a number of such field studies have been carried out and can be used to "calibrate" laboratory-derived predictions of risk based on acute endpoints. Laboratory data are most useful in providing a comparative assessment of the risk posed by different pesticides. Such comparative assessments have also proved useful in the various measurement systems designed to assess the environmental consequences of choosing different agrochemicals or agricultural management systems (Benbrook et al. 1996).

When carrying out comparative risk assessments for pesticides, it is essential to use the most unbiased data possible. Pesticides are customarily tested against no more than 1 to 3 bird species, yet there are an estimated 10,000 species in the world. More than 800 species occur in Canada and the United States alone. In this review, we present acute toxicity values that can be used as reference values in pesticide risk assessments. These values are only useful for protecting birds from pesticides if matched by adequate measures of exposure. Also, they only address acute lethal toxicity and not reproductive or chronic health effects or even sublethal effects that may give rise to delayed mortality or a reduction in biological fitness. Different strategies have been used over the years to compare the toxicity of different pesticides to birds. (1) Restricting among-chemical comparisons to a commonly tested group of species: This strategy quickly runs into data gaps and leads to an arbitrary ranking of relative toxicity depending on the species chosen. Test species can be inconsistent in their relative sensitivity rankings among pesticides (Tucker and Haegele 1971). (2) Several species as phylogenetically close as possible to a species of interest are used: Unfortunately, toxicological susceptibility does not always follow phylogenetic lines (Schafer and Brunton 1979; Mineau 1991; Joermann 1991; Baril et al. 1994), and it is notoriously difficult to predict which species are most at risk from a given pesticide treatment. (3) Finally, using the lowest value available from all species tested: This approach, however, introduces a systematic bias related to the amount of test data available and reported for each pesticide.

Baril et al. (1994) and then Luttik and Aldenberg (1995, 1997) suggested that a distribution-based method should be employed for birds much as had been proposed for soil and aquatic organisms (Stephan and Rogers 1985; Kooijman 1987; Van Straalen and Denneman 1989). A distribution-based approach uses the pesticide-specific data available to define the shape of the distribution through the estimation of a mean and variance for the distribution. Fitting toxicity data to distributions has been criticized, most recently by Newman et al. (2000). However, their main criticism concerns the pooling of toxicity data for very different phylogenetic groups and the resulting lack of fit to commonly used distributions. This result would be expected, for instance, when pooling the response of both algae and aquatic invertebrates to a herbicide. We have found that both the log-logistic and log-normal distributions are adequate when dealing with the toxicity of pesticides to birds. The alternative proposed by Newman and colleagues is to use bootstrapping (distribution-free resampling of the data) to arrive at an effect threshold. The main disadvantage of this technique is that any biases inherent in the initial data, a likely problem if few data are available, will be preserved and emphasized. We therefore opted for a distribution-based method. However, two modifications of this technique were necessary: (1) introducing a scaling factor for body weight to improve cross-species comparisons of toxicological susceptibility (Mineau et al. 1996); and (2) developing a strategy to consider chemicals for which there are insufficient data from which to derive a distribution. Slightly different approaches were described by Baril et al. (1994) and by Luttik and Aldenberg (1995) to meet this second objective. We present here the relative merits of both these approaches, and provide what we believe to be the most scientifically defensible reference values that can be used for assessing the *relative* acute risk of different pesticides to birds.

## II. Data Selection
### A. Data Acquisition and Appraisal

Data acquisition procedures were modified slightly from those described in Baril et al. (1994) and Mineau et al. (1996). Under the auspices of the OECD (Organisation for Economic Co-operation and Development) and following the recommendations of a 1994 workshop on avian toxicity testing (SETAC 1996), the Canadian Wildlife Service's existing database of $LD_{50}$ values was expanded with the assistance of several collaborators worldwide. A recent (Brian J. Montague 30 April 98 pers. comm.) version of the U.S. EPA 'one liner' database was obtained and amalgamated with our existing database. Between 1996 and 1997, various data were also obtained from Germany (Federal Biological Research Centre for Agriculture and Forestry), the Netherlands (National Institute of Public Health and the Environment), France (Institut National de la Recherche Agronomique), and the United Kingdom (Pesticide Disclosure Documents from the Pesticide Safety Directorate). Many of those data consisted of studies sponsored by pesticide manufacturers in support of the registration of their pest control products. The confidentiality afforded to these data varies greatly between countries. At one extreme (Canada), all data endpoints (e.g., $LD_{50}$ values) are considered to be proprietary and confidential unless specifically marked for public release by the manufacturer; at the other extreme (U.S.), endpoints are freely available and complete $LD_{50}$ studies can be requested through "Freedom of Information Act" provisions. Some jurisdictions make endpoints available for a fee, either in hard copy form (United Kingdom) or through a dial-up database (France). Furthermore, several companies release data endpoints in summary form (e.g., Material Safety Data Sheets) and these data are picked up by public sources such as the British Crop Protection Council's *Pesticide Manual* (see following). Because of this confusion, we decided that no species-specific data would be released nor could any be back-calculated from the information presented here.

Known sources of data from the open literature were also searched. These sources included existing compendia of avian acute toxicity data assembled by governmental agencies in the U.S. and elsewhere (Schafer et al. 1983; Hudson et al. 1984; Grolleau and Caritez 1986; Smith 1987), as well as scientific publications containing one or a few values. An exhaustive search of the literature was carried out by means of the Terretox database of the U.S. government as well as the commercially available Medline and Biological Abstracts databases. A final source of data consisted of various editions of the *Pesticide Manual* (the 4th, 9th, and 11th editions; the latter one containing most of the avian data). Before accepting the data from these editions of the *Pesticide Manual*, a sample was checked against our existing database. Although the data presented in the *Pesticide Manual* (Tomlin 1997) were often biased toward species of lesser sensitivity, the data themselves were almost invariably accurate (a calculated accuracy rate of 99% for 108 data points for which the original data source was already available to us). The 11th edition of the *Pesticide Manual* was also used to indicate which pesticides are currently being commercialized worldwide. All data were carefully vetted for errors and duplicates and, where possible, checked to the original source.

## B. Selection and Processing of Toxicity Data

The distribution approach to handling interspecies differences in sensitivity to chemicals requires that a single value be available for each pesticide-species combination. In many cases, however, more than one value was available for any combination. It was therefore necessary to establish criteria that provided a uniform process for the selection of values. The criteria were chosen so as to minimize bias and variability introduced by the formulation of the pesticide, the age of the birds, and numerous other factors. Our data selection criteria were modified to agree with those used by Luttik and Aldenberg (1995, 1997) following several rounds of consultation. This method explains some of the small discrepancies between the values reported here and the few that were presented in Mineau et al. (1996). The criteria were as follows.

1. Only Data for Adult-Sized Birds Were Used (typically >1 month for passerine and gallinaceous birds; >3 months for waterfowl). In some cases, age was unspecified but the data, often generated for pesticide submissions, were assumed to refer to adults as specified in current and former EPA protocols. Tests on passerine species were generally carried out on wild-caught individuals which, we assumed, had fledged at that point. The notable exception was the domestic chicken, for which it is customary to test young chicks or pullets. Values for domestic fowl were therefore excluded unless age was specified.

2. Studies of formulated products or of technical products with very low percentages of active ingredient were not used. We did not correct for the percentage of active ingredient found in the technical grade of the pesticide.

3. If more than one $LD_{50}$ value was available for a species and a given pesticide, a geometric mean value was calculated. No one study was given preference over another.

4. If there were multiple $LD_{50}$ values for a certain species and a given pesticide, and one of those values was a "greater than" (>) or "lower than" (<) value, this value was not used if it lay inside the range of the other available values. However, this value was used as a point estimate (the < or > having been removed) when it lay outside the range of other available values for that species and pesticide.

5. If, in a set of available $LD_{50}$s for a given pesticide, a particular species only had a greater or lower than value, this value was not used if it lay within the range of values available for the other species, but it was used as a point estimate (the < or > having been removed) if it were outside the range of values available for the other species.

6. Where the majority of available species $LD_{50}$s were greater than a certain value, such as is often the case with nontoxic pesticides where limit values are given (e.g., all species >2000 mg/kg), the data were not fitted to a distribution regardless of how many such values were available. Instead, an extrapolation factor approach was employed, as described next.

7. When separate values were given for each sex, the geometric mean of the two values was calculated.

8. When the value given by a single source was a range, the geometric mean of the range was calculated. This criterion applied mainly to the studies reported by Schafer and coauthors (1983) in which the ranges given correspond to values obtained from separate studies (Ed Shafer, personal communication).

9. When compendia of values were published by any given laboratory and where there were discrepancies between different editions or publications, the most recently published value was accepted. This choice assumes that previous errors were corrected by the laboratory in question.

Unfortunately, we were not able to take into account the method of dosing (e.g., by gavage needle or gelatin capsule) nor were we able to account for the use of vehicles or diluents (e.g., corn oil), this information seldom being available. We recognize this is an important source of variation as is the differential propensity of different bird species to regurgitate (Hart and Thompson 1995). Also, we were unable to ensure that the technical pesticide material was identical from test to test. The data span several decades, and it is likely that the level of purity and proportion of degradation products and contaminants changed over the years and across different manufacturers. Some pesticides are known to be racemic mixtures and may have been marketed first as the mixture and later as the active isomer. Often, this is reflected by the fact that there are several CAS numbers available for any one pesticide (Tomlin 1997). We cannot be certain that all toxicity tests reported here are specific to the CAS number given for the pesticide in question, nor can we venture a guess as to the contribution of this factor to the overall within-chemical variance in toxic response.

As with any such compendium, cholinesterase-inhibiting pesticides are well represented (at least in having higher numbers of species tested per pesticide) because of their relatively high toxicity to birds and the fact that they account for the majority of wildlife poisoning incidents. Because of our desire to develop methods that are representative of all classes of pesticides, we carried out some analyses (at least initially) on cholinesterase inhibitors separately from other pest control products. The data for cholinesterase inhibitors are therefore presented separately in this review (see tables 2 and 3). The database thus compiled for cholinesterase inhibitors consists of 147 pesticides and 837 acceptable $LD_{50}$ determinations. For noncholinesterase inhibitors, we were able to compile 1601 acceptable $LD_{50}$ values for 733 pesticides.

## C. Scaling of Toxicity Data to Body Weight

It is customary to extrapolate between species on the basis of acute toxicity measurements expressed in mg/kg body weight, yet Mineau et al. (1996) showed that for a group of 36 pesticides chosen for the high number of $LD_{50}$ data points available, the appropriate scaling factor in birds (with toxicity in mg/animal regressed against body weight) is usually > 1 and can be as high as 1.55. These authors showed that, when fitting a distribution to $LD_{50}$ data expressed as mg/kg body weight (i.e., forcing the data to a slope of 1), the resulting distribution overestimated $LD_{50}$ values for small-bodied birds and resulted in wider confidence intervals for the usual distribution-based toxicity benchmarks, e.g., the 5% and 95% bounds of the distribution. For the present analysis, and for all pesticides with $n \geq 4$, ln $LD_{50}$ values (in mg/kg rather than mg/animal) were regressed against ln weight in grams. Mean species weights were obtained from Dunning (1993). As described by Mineau et al. (1996), the slope or "scaling factor" for the majority of pesticides is positive [corresponding to a slope greater than 1 in Mineau et al. (1996) when toxicity values were expressed as mg/animal]. For the 130 pesticides with $n \geq 4$, the regressions were positive in 99 cases and the overall mean slope was 0.239; this was slightly more extreme than the value of 1.15 (corresponding to 0.15 when expressed on the basis of mg/kg) reported by Mineau et al. for a subset of 36 pesticides. A similar proportion of slopes was positive for cholinesterase inhibitors (54/68) than for other pesticides combined (45/62). As determined by an $F$-test, 14 pesticides of the 130 had a slope significantly different from 0 at the $p < 0.05$ probability level (11 of which were positive). Allowing the $p$ to rise to 0.1, a total of 30 slopes were significantly different from null, 24 of them positive. As argued by Mineau et al. (1996), failure to achieve statistical significance in a majority of the slopes does not remove the biological significance of the finding. Several slopes may miss being significantly different from 0 either because of inadequate sample size, or because they are only slightly different from 0, or a combination of both. The fact that the large majority are greater than 0 indicates a phenomenon that is biologically significant. Therefore, even for pesticides for which statistical significance is not achieved, the observed slope should be regarded as a better estimate of the true value rather than assuming the slope is 0. Not using a scaling factor when susceptibility does scale to weight would result in potentially serious underprotection at one end (usually the small one) of avian size ranges. On the other hand, scaling toxicity data to a completely spurious slope factor might mislead. Fortunately, inspection of the data revealed that many of the very extreme slopes measured, both positive and negative, were in fact significant, at least at the $p < 0.1$ level, thus lessening the concern that we may be fitting data to spurious slopes.

The reason for avian acute toxicity values scaling to weight raised to the value of 1.2 or even slightly higher on average is unclear. In mammals, the common wisdom is that toxicity tends to scale to 0.67 or 0.75 (reviewed in Mineau et al. 1996) although a recent reassessment of acute toxicity data in mammals and birds (Sample and Arenal 1999) calculated an average scaling factor of 0.94 for mammals and confirmed an average of 1.2 in birds. It was suggested (Fischer and Hancock 1997) that these scaling factors may be a consequence of taxonomic differences. Songbirds (order Passeriformes) constitute the majority of the small-bodied species in the database, and they may be more susceptible toxicologically. A simpler, and more compelling possibility, is that small birds in any taxonomic group are less able to withstand the rigors of the physiological disruption brought about by acute dosing, especially reduced food intake. Regardless of the reason, the end result is the same: the use of the appropriate scaling factor results in the least estimation error for the toxicity of a pesticide to a bird of a given weight and in reduced variance in the distribution of $LD_{50}$ values. One, however, needs to exercise caution in estimating $LD_{50}$ values for very large or very small bird species, or birds from taxonomic groups that are poorly represented in the data set. For example, as reviewed by Mineau et al. (1999) there is some evidence that hawk and owl species (orders Falconiformes and Strigiformes) are more sensitive, at least to cholinesterase inhibitors, than other birds of similar body weight.

## D. Modification of the Distribution Approach to Incorporate Body-Weight Scaling

Until now the best method available to derive a set level of protection with a given level of certainty based on the distribution of toxicity data was that developed by Aldenberg and Slob (1993). These authors modified existing methods (Kooijman 1987; Van Straalen and Denneman 1989; Wagner and Lokke 1991), which aimed to determine environmental concentrations or doses of chemicals that were protective of 95% of the species in the wild. The modifications consisted of deriving extrapolation constants, $K_n$, that account for the uncertainty in the estimates of the distribution parameters when dealing with small sets of laboratory-derived toxicity data. The extrapolation constants are determined such that a one-sided left confidence limit $L$ for $\log(HD_5)$ (for Hazardous Dose at the 5% tail of the distribution) is given by

$$L = \bar{y}_n - [K_n * S_n]  \tag{1}$$

where $\bar{y}_n$ and $S_n$ are mean and standard deviation, respectively, of a sample log ($LD_{50}$) test data of size $n$.

This approach does not, however, incorporate body weight as a covariate. Mineau et al. (1996) were able to describe the relationship between the $LD_{50}$ and body weight using the equation

$$y_i = \beta_0 + \beta_1 x_i + e_i  \tag{2}$$

where, $y_i = \log(LD_{50})$, $x_i = \log(Weight)$, $\beta_0$ and $\beta_1$ denote the intercept and slope of the regression line, and $e_i$ denotes the random deviation of the data from the model.

The distributional method as described by Aldenberg and Slob (1993) does not incorporate a body weight covariate and requires an estimate of the mean and the standard deviation or variance. The equivalent terms can be defined for a model with a covariate, but in this case one must choose a value for $x_0$ (the log of the weight of the bird that one wants to protect). The estimators of these two quantities with and without a covariate are compared in Appendix 1. For the Aldenberg and Slob approach, the precision of the estimates only depends on $n$ whereas for the covariate approach it depends on $n$ and another term that reflects the precision of the predicted $LD_{50}$ at the designated weight. The predicted $LD_{50}$ becomes less precise as one moves away from the average of the $x_i$. To compensate for this, a value for the extrapolation constant $K$ must be developed separately for each data set and designated weight of bird to be protected; this is done through a simulation in the manner of Aldenberg and Slob (1993).

### E. Choice of an Appropriate Percentile in the Acute Toxicity Distribution

Working with a distribution allows one to set a desired percentile, or a threshold $LD_{50}$ value sufficiently protective for an arbitrarily chosen proportion of the entire population of bird species. For risk assessment purposes, it is customary to choose a value in the left tail, e.g., at the 1% or 5% tail of the distribution. This custom was born of convenience and practicality and has no scientific basis. Some may take issue with the fact that 5% or even 1% of the species inhabiting the exposed ecosystem can be summarily "written off." Choosing a percentile does not mean that this percentage of species will necessarily be impacted. The final level of protection afforded to birds will depend on the interplay of all the components of the risk assessment and regulatory scheme. For comparative risk assessment, using an $LD_{50}$ value representative of the left tail of the distribution is preferable to using a measure of central tendency (median, arithmetic mean or geometric mean) because the latter parameters do not take into account information regarding the variance of the distribution. For the purpose of this exercise, the 5th percentile of the log-logistic distribution of species $LD_{50}$s was determined (the point on the tail of the log-logistic distribution that excludes 5% of the lower $LD_{50}$ values). It is useful to also remember that we are fitting $LD_{50}$ values to a distribution; these are not no-effect or minimal-effect concentrations. At the predicted threshold, half the exposed individuals from a species at the 5% tail of susceptibility are expected to die. Risk assessors may wish to apply another factor to cover intraspecific differences in susceptibility. Typically, this would be done through consideration of the probit slope of the dose–mortality relationship.

Having arbitrarily fixed the protection level at the 5th percentile of the species distribution, (termed $HD_5$ (Hazardous Dose 5%) by Aldenberg and colleagues as well as in this review, or $TLD_5$ (Threshold Lethal Dose 5%) by Baril and colleagues), we still need to fix the level of certainty we wish to attach to the determination of this value. As argued by Aldenberg and Slob (1993) and confirmed by Baril et al. (1994), median estimates of the $HD_5$ (calculated with a 50% probability of overestimation) may not be sufficiently protective, especially where rare or valuable focal species of unknown susceptibility must be protected. In other words, far fewer than 95% of bird species may actually be protected. However, thresholds that account for a higher (90% or 95% are commonly used) certainty that the estimate of the 5th percentile is not overestimated are extremely conservative (likely overprotective), especially when sample sizes are small. Our intent here was to present threshold values that would allow for the "fairest" comparison possible between pesticides having data sets of vastly different size and quality. Indeed, one of our goals was to permit comparison of older pesticides (often with large data sets) with that of newer products (with typically few data) without allowing sample size to have an overwhelming influence on the result. We opted therefore to report median threshold values (the 5% tail of the distribution calculated with 50% probability of overestimation), recognizing that some of these values carry with them a very high risk of underprotection, which may render them unsuitable for product by product risk assessment. Of course, if probabilistic risk assessments are to be carried out, one may wish to enter the entire estimated distribution of $LD_{50}$ values rather than an arbitrary 5% threshold.

## F. Choice of the Appropriate Bird Weights to Model

Because of the nature of regression, the real advantage of using the new distribution model that incorporates body weight as a covariate is predicting the sensitivity of birds having body weights that deviate from the tested average. For chemicals for which the regression between body weight and toxicity is highly statistically significant, the confidence in the estimated $HD_5(50\%)$ for birds of any chosen weight is high. For those chemicals for which the slope is different from zero but not statistically significant, we opted to apply the covariate as well. Using a distribution model without a covariate can lead to the erroneous underestimation of the sensitivity of birds at either extreme of the weight axis if the scaling relationship is real. Because our intent was to provide reference values that are sufficiently protective of most birds regardless of size, we assumed that the fitted scaling relationship was biologically real in all cases.

To generate a single $LD_{50}$ value protective of birds in general regardless of their weight, we first calculated $HD_5$ values for birds of 20, 100, 200, or 1000 g. Those weights were chosen based on typical bird weights recorded from casualties in pesticide field studies carried out in North America, primarily the U.S. (CWS unpublished analysis). Although this weight range does not encompass all bird species, it does account for the vast majority. The reference value reported in this review is the lesser of the 5% threshold values calculated for these four body weights. Most often, this was the value determined for birds of 20 g.

### G. Derivation of Extrapolation Factors

As proposed by Luttik and Aldenberg (1997), we decided that a minimum of four $LD_{50}$ values were needed for determining the mean and standard deviation parameters of the distribution. For pesticides with fewer data points, a different approach is needed because the data are insufficient to fit to a distribution without the parameters being estimated with unusually large errors. We must therefore predict the fifth percentile of the distribution through extrapolation from the small data set using some predetermined factor or set of factors. Luttik and Aldenberg (1995) presented one approach to derive such extrapolation factors.[1] Their method assumes the following: (1) the mean of the logarithm of the available toxicity data is the best estimate of the mean of the distribution; (2) the standard deviation is equal to a "generic" standard deviation that is calculated from the pooled historical data sets for a large number of chemicals tested on many species; and (3) species sensitivities are random across chemicals. Thus, Luttik and Aldenberg determined that, if a single test species is available, an extrapolation factor of 5.7 should be applied to the $LD_{50}$ (unadjusted for body weight scaling) to obtain the median estimated $HD_5$ (to be indicated as $HD'_5(50\%)$). This factor stays constant regardless of $N$, the number of species for which test data are available.

There is, however, a serious impediment to the use of such a strategy for regulatory or comparative purposes. Data available for any given pesticide (especially if the data are limited to one or a few species) are not necessarily a random sample. As discussed earlier, we have found that only a few data points are typically released, and these data may be biased to put a product in the best light possible regarding its toxicity to nontarget species. Also, species commonly used for testing (e.g., the mallard duck and northern bobwhite) tend to be at the "insensitive" side of the distribution already. We believe that to adopt the strategy of a single "universal" extrapolation factor would be a strong incentive to biased reporting. Several authors have found that species tend to differ

---

[1]The word extrapolation factor is used here in place of the frequently used safety factor or assessment factor. In contrast to the latter, which often are arbitrarily set at 10 or 100 to reflect a potentially error-prone extrapolation, the extrapolation factors presented here are based on sound empirical data reflecting variability in interspecies susceptibility to chemicals.

in their sensitivity to pesticides in a predictable manner; some species are, on average, more sensitive than others to pesticides (Baril et al. 1994; Joermann 1991; Schafer and Brunton 1979). To make use of these known relationships, Baril and colleagues proposed that extrapolation factors should be tailored to the actual species for which data are available. For example, a different extrapolation factor would be applied to a single mallard $LD_{50}$ value than to a northern bobwhite or Japanese quail $LD_{50}$. A simulation exercise (Baril and Mineau 1996; this is an abstract only; data not published) showed that to ignore sensitivity relationships in favor of a single (universal) extrapolation factor resulted in more estimation errors.

We therefore opted to generate species-specific extrapolation factors from our available data set. The calculated $HD_5(50\%)$ values for all pesticides with $N$ greater or equal to 6 were used to derive extrapolation factors. All computed $HD_5(50\%)$ values were used regardless of the significance of the regression statistics. As outlined earlier, the smaller of the $HD_5$ values computed for weights ranging between 20 and 1000 g was retained as our reference value, and those reference values were used to calculate species- and pesticide-specific extrapolation factors. A sample size of 6 or more ensured that both the parameters of the distribution and the slope of the regression between body weight and $LD_{50}$ were relatively well characterized. Thus, the extrapolation factor specific to a particular species or combination of test species is simply the ratios of the computed $HD_5(50\%)$ to the $LD_{50}$s of the test species (or geometric mean of $LD_{50}$ values in the case of a combination of species), averaged over all chemicals in the database.

We propose that these extrapolation factors be used to estimate $HD_5(50\%)$ values ($HD_5(50\%)$) where $N$, the number of species tested is, 1–3. Two different methods of combining the two or three available $LD_{50}$ values were tried: taking the smallest value, and taking the geometric mean of the values. Deriving extrapolation factors from the geometric means of available $LD_{50}$s was found to lead to less estimation error (data not shown). The geometric mean was then retained as the best way to combine two or three available data points to compute extrapolation factors. We computed extrapolation factors for most of the commonly tested species or combinations of those species (Table 1). Other factors computed for more rarely-tested species were derived as needed but are not given here because of the smaller sample sizes used in their derivation.

The use of an extrapolation factor introduces another potentially significant source of error in the estimation of the $HD_5(50\%)$; this is the error resulting from an individual species' varying sensitivity relative to the population 5% tail. Whereas some species appear to be reasonably "well behaved" by showing a degree of sensitivity relative to the population tail that is consistent across chemicals (red-winged blackbird, red-billed quelea, Japanese quail), other species tended to move extensively within the sensitivity distributions established for each pesticide (chicken, mallard, European starling). For the latter species, sometimes they were very sensitive and sometimes they were very insensitive relative to the 5% percentile. The extrapolation factors were therefore ranked

Table 1. Extrapolation factors ordered by increasing coefficients of variation.

| | $n$ | Extrapolation factor | Approximate C.V. | 95th percentile | 5th percentile |
|---|---|---|---|---|---|
| Red-winged blackbird | 67.00 | 3.95 | 2.32 | 10.62 | 1.47 |
| Red-billed quelea | 22.00 | 3.63 | 2.68 | 10.92 | 1.20 |
| Bobwhite quail and Japanese quail | 36.00 | 8.05 | 3.18 | 27.97 | 2.32 |
| Bobwhite quail and Japanese quail and mallard duck | 32.00 | 8.94 | 3.64 | 34.85 | 2.29 |
| Japanese quail | 61.00 | 10.36 | 4.11 | 44.96 | 2.39 |
| Japanese quail and mallard duck and house sparrow | 46.00 | 8.37 | 4.26 | 37.51 | 1.87 |
| Japanese quail and mallard duck | 56.00 | 10.41 | 5.48 | 58.93 | 1.84 |
| European starling and red-winged blackbird | 57.00 | 5.63 | 5.57 | 32.37 | 0.98 |
| Bobwhite quail and mallard duck | 40.00 | 9.61 | 6.10 | 60.12 | 1.54 |
| Bobwhite quail | 42.00 | 8.61 | 7.19 | 63.08 | 1.17 |
| Ring-necked pheasant | 66.00 | 9.41 | 7.42 | 71.05 | 1.25 |
| European starling | 59.00 | 11.82 | 7.60 | 91.32 | 1.53 |
| Mallard duck | 67.00 | 10.38 | 10.89 | 114.00 | 0.95 |
| Chicken | 37.00 | 19.75 | 13.82 | 274.31 | 1.42 |

Approx. C.V. = (95th percentile − 5th percentile)/extrapolation factor.
95th percentile = extrapolation factor + 1.645 * SD.
5th percentile = extrapolation factor − 1.645 * SD.

on the basis of their approximate coefficient of variation, the lowest coefficient being indicative of the lowest likelihood of making a large estimation error. Given the possibility of applying more than one extrapolation factor, we chose the one with the lowest approximate coefficient of variation. For example, although an extrapolation factor based on the geometric mean of several species is usually more "stable" and therefore less prone to serious error than a factor based on a single species, it can be seen from Table 1 that extrapolating the $HD'_5(50\%)$ from a single Japanese quail $LD_{50}$ value results in less error, on average, than extrapolating from a combination of bobwhite and mallard data. Nevertheless, using an extrapolation factor is clearly a "second-best" alternative to curve-fitting $LD_{50}$ data and deriving an actual $HD_5$. In the example just given, testing two species such as the northern bobwhite and mallard increases the probability that unexpected chance variations in susceptibility will be uncovered.

We believe that the approach of using reference values based on species-specific extrapolation factors represents the most unbiased attempt to date to compare the toxicity of pesticides for which many data points are available with those about which we know very little. Because cholinesterase-inhibiting pesticides represent a large group of chemicals with a uniform mode of toxic action, they were analyzed separately from other pesticides, and specific extrapolation factors were derived for that group of compounds. However, when the comparison was made between those compounds and a sample of noncholinesterase inhibitors, we found that the computed mean extrapolation factors were not significantly different (not shown) and a single set of factors for all pesticides in the sample was therefore generated (see Table 1).

## H. Procedure for Pesticides Having Low Acute Toxicity

In the case of pesticides of lower toxicity, limit values are often provided; e.g., $LD_{50} > 2000$ mg/kg. Ideally, one would like to take into consideration the condition of the test birds and any mortality at the limit dose. In trying to set a value that is protective of 95% of bird species, it matters whether birds at the limit dose were moribund, with possibly some mortality being seen already, or whether there were no visible signs of toxicity in any of the individuals tested. Unfortunately, this information was not uniformly available. The following compromise procedure was therefore developed. (1) If, for any pesticide data set, a toxicity data point with an exact value existed (rather than a limit), this data point was used with the relevant species-specific extrapolation factor where possible. (2) Otherwise, the relevant species-specific extrapolation factors were applied separately to each available data point (treated as point estimates, ignoring the > symbol) and the *highest* resulting $HD'_5$ value was retained. Even then, it is clear that the resulting $HD'_5$ is probably an underestimate and that the pesticide is likely less toxic than shown. However, given that the $HD'_5$ thus calculated is already very high, this underestimation is unimportant in the context of a relative risk assessment. The acute toxicity of these pesticides is not likely to be a concern.

Table 2. Reference LD$_{50}$ values for Cholinesterase-inhibiting pesticides ordered alphabetically by common chemical name.

| Compound | CAS_RN | n | Median | Slope | Intercept | p | HD$_5$(50%) | #_EF | Status |
|---|---|---|---|---|---|---|---|---|---|
| Acephate | 30560-19-1 | 7 | 146.00 | 0.1809 | 4.7716 | 0.66 | 18.52 | — | E |
| AKTON™ | 1757-18-2 | 2 | 1037.50 | na | na | na | 18.99 | 1 | U |
| Aldicarb | 116-06-3 | 10 | 2.82 | 0.2955 | -0.6559 | 0.12 | 0.43 | — | E |
| Allyxycarb | 6392-46-7 | 2 | 12.42 | na | na | na | 3.37 | 1 | S |
| Aminocarb | 2032-59-9 | 4 | 46.20 | -0.3618 | 6.0268 | 0.33 | 6.59 | — | S |
| Azamethiphos | 35575-96-3 | 2 | 39.30 | na | na | na | 3.98 | 2 | E |
| Azinphos-ethyl | 2642-71-9 | 1 | — | na | na | na | 1.53 | 1 | E |
| Azinphos-methyl | 86-50-0 | 7 | 44.69 | 0.7806 | -0.6793 | 0.00 | 2.28 | — | E |
| Bendiocarb | 22781-23-3 | 4 | 16.24 | -0.9475 | 8.3724 | 0.37 | 0.72 | 1 | E |
| Benfuracarb | 82560-54-1 | 1 | — | na | na | na | 4.23 | 1 | E |
| BOMYL™ | 122-10-1 | 2 | 5.36 | na | na | na | 0.25 | 1 | U |
| Bromophos | 2104-96-3 | 1 | — | na | na | na | 491.14 | 1 | S |
| Bromophos-ethyl | 4824-78-6 | 5 | 300.00 | 0.8503 | 0.805 | 0.01 | 12.88 | — | E |
| Bufencarb | 8065-36-9 | 8 | 32.95 | 0.1165 | 2.7226 | 0.71 | 3.09 | — | S |
| Butocarboxim | 34681-10-2 | 1 | — | na | na | na | 6.17 | 1 | E |
| Butonate | 126-22-7 | 3 | 158.00 | na | na | na | 40.00 | 1 | S |
| Butoxycarboxim | 34681-23-7 | 1 | — | na | na | na | 18.58 | 1 | E |
| Cadusafos | 95465-99-9 | 2 | 123.05 | na | na | na | 6.33 | 2 | E |
| Carbanolate | 671-04-5 | 12 | 4.22 | 0.1412 | 0.8936 | 0.53 | 0.75 | — | S |
| Carbaryl | 63-25-2 | 7 | 1870.50 | 0.8026 | 2.5147 | 0.06 | 30.05 | — | E |
| Carbofuran | 1563-66-2 | 18 | 1.65 | 0.0423 | 0.257 | 0.82 | 0.21 | — | E |
| Carbophenothion | 786-19-6 | 9 | 56.80 | 0.4054 | 1.431 | 0.12 | 2.00 | — | S |
| Carbosulfan | 55285-14-8 | 2 | 51.00 | na | na | na | 9.52 | 1 | E |

Also tabled are the CAS registration numbers, the number of species values available, the median LD$_{50}$, the slope, intercept and p value of the LD$_{50}$ *weight regression, the calculated or estimated HD$_5$(50%) and, where used, the number of species from which an extrapolation factor was developed (see Table 1). The status follows the nomenclature of the *Pesticide Manual*, 11th Ed.; E, in use; S, superseded; U, unknown.

Table 2. (Continued).

| Compound | CAS_RN | n | Median | Slope | Intercept | p | HD$_5$(50%) | #_EF | Status |
|---|---|---|---|---|---|---|---|---|---|
| Chlorethoxyfos | 54593-83-8 | 1 | — | na | na | na | 3.25 | 1 | E |
| Chlorfenvinphos | 470-90-6 | 15 | 23.70 | 0.3635 | 1.8073 | 0.10 | 2.73 | — | E |
| Chlormephos | 24934-91-6 | 3 | 100.00 | na | na | na | 25.10 | 1 | E |
| Chlorphoxim | 14816-20-7 | 2 | 100.00 | na | na | na | 11.61 | — | S |
| Chlorpyrifos | 2921-88-2 | 18 | 27.36 | 0.226 | 2.0941 | 0.09 | 3.76 | 1 | E |
| Chlorpyrifos methyl | 5598-13-0 | 2 | 845.00 | na | na | na | 25.32 | 1 | E |
| Chlorthion | 500-28-7 | 3 | 280.00 | na | na | na | 70.89 | 1 | S |
| Cloethocarb (Bas 2631) | 51487-69-5 | 1 | — | na | na | na | 0.43 | 1 | S |
| Coumaphos | 56-72-4 | 12 | 6.78 | 0.2179 | 0.7579 | 0.36 | 0.69 | — | E |
| Crotoxyphos | 7700-17-6 | 2 | 423.10 | na | na | na | 14.23 | 1 | S |
| Crufomate | 299-86-5 | 2 | 182.50 | na | na | na | 25.32 | 1 | S |
| Cyanophos | 2636-26-2 | 1 | — | na | na | na | 0.83 | 1 | E |
| Demeton | 8065-48-3 | 13 | 7.67 | 0.2357 | 0.6881 | 0.15 | 1.04 | — | S |
| Demeton-S-methyl | 867-27-6 | 7 | 49.00 | 0.1396 | 3.0132 | 0.47 | 7.24 | — | E |
| Demeton-S-methylsulphon | 17040-19-6 | 2 | 50.06 | na | na | na | 8.14 | 1 | S |
| Dimidafos | 1754-58-1 | 2 | 44.15 | na | na | na | 3.37 | 1 | S |
| Diazinon | 333-41-5 | 14 | 5.25 | -0.2608 | 3.5883 | 0.29 | 0.59 | — | E |
| Dicapthon | 2463-84-5 | 1 | — | na | na | na | 4.13 | 1 | S |
| Dichlofenthion | 97-17-6 | 7 | 75.00 | 0.2685 | 3.327 | 0.57 | 7.54 | — | S |
| Dichlorvos (DDVP) | 62-73-7 | 11 | 14.75 | 0.0493 | 2.2876 | 0.61 | 5.18 | — | E |
| Dicrotophos | 141-66-2 | 15 | 2.83 | 0.1787 | 0.3645 | 0.32 | 0.42 | — | E |
| Diethofencarb | 87130-20-9 | 2 | 2250.00 | na | na | na | 234.13 | 2 | E |
| Dimethoate | 60-51-5 | 10 | 29.50 | 0.1773 | 2.5296 | 0.37 | 5.78 | — | E |
| Dimetilan | 644-64-4 | 4 | 27.20 | 0.1601 | 2.8008 | 0.83 | 0.92 | — | S |
| Dioxacarb | 6988-21-2 | 5 | 115.00 | 0.7849 | 0.4173 | 0.09 | 3.36 | 1 | S |
| Dioxathion | 78-34-2 | 2 | 258.50 | na | na | na | 25.50 | — | S |
| Disulfoton | 298-04-4 | 7 | 11.90 | 0.2019 | 1.2104 | 0.60 | 0.81 | — | E |

Table 2. (Continued).

| Compound | CAS_RN | n | Median | Slope | Intercept | p | HD$_5$(50%) | #_EF | Status |
|---|---|---|---|---|---|---|---|---|---|
| EPN | 2104-64-5 | 14 | 6.43 | 0.3624 | 0.604 | 0.33 | 0.53 | — | E |
| Ethamphos (ethamfenphos, ethamfenthion) | | 2 | 7.93 | na | na | na | 0.78 | 2 | U |
| Ethiofencarb | 29973-13-5 | 3 | 196.00 | na | na | na | 14.96 | 1 | E |
| Ethion (diethion) | 563-12-2 | 5 | 127.80 | 0.9169 | -0.1551 | 0.18 | 1.06 | — | E |
| Ethoprop | 13194-48-4 | 9 | 7.50 | 0.1086 | 1.3764 | 0.40 | 2.41 | — | E |
| Etrimfos | 38260-54-7 | 1 | — | na | na | na | 23.65 | 1 | E |
| Famphur | 52-85-7 | 3 | 2.70 | na | na | na | 0.45 | 1 | E |
| Fenamiphos | 22224-92-6 | 5 | 1.10 | -0.0863 | 0.5444 | 0.66 | 0.43 | — | E |
| Fenchlorphos | 299-84-3 | 4 | 487.42 | 0.6983 | 2.1557 | 0.15 | 12.23 | — | S |
| Fenitrothion | 122-14-5 | 12 | 63.43 | 0.2707 | 3.0869 | 0.39 | 3.37 | — | E |
| Fenobucarb | 3766-81-2 | 1 | — | na | na | na | 31.12 | 1 | E |
| Fensulfothion | 115-90-2 | 14 | 0.73 | 0.2903 | -1.8278 | 0.05 | 0.13 | — | S |
| Fenthion | 55-38-9 | 23 | 5.62 | 0.2581 | 0.4784 | 0.07 | 0.87 | — | E |
| Fonofos | 944-22-9 | 10 | 23.50 | 0.3442 | 1.293 | 0.07 | 3.86 | — | E |
| Formetanate | 22259-30-9 | 4 | 31.75 | -0.515 | 6.2418 | 0.06 | 8.77 | — | E |
| Fospirate | 5598-52-7 | 2 | 34.75 | na | na | na | 3.37 | 1 | S |
| Fosthiazate | 98886-44-3 | 2 | 15.00 | na | na | na | 1.47 | 2 | E |
| Furathiocarb | 65907-30-4 | 1 | — | na | na | na | 2.41 | 1 | E |
| Heptenophos | 23560-59-0 | 2 | 52.74 | na | na | na | 3.04 | 1 | E |
| Isazofos | 42509-80-8 | 3 | 11.10 | na | na | na | 0.51 | 2 | E |
| Isocarbophos | 24353-61-5 | 5 | 1.00 | 0.254 | -1.1291 | 0.148 | 0.26 | — | U |
| Isofenphos | 25311-71-1 | 6 | 10.96 | 0.0994 | 2.7255 | 0.86 | 0.44 | — | E |
| Isoprocarb | 2631-40-5 | 1 | — | na | na | na | 14.23 | 1 | E |
| Leptophos | 21609-90-5 | 5 | 268.84 | 1.2259 | -1.5553 | 0.41 | 0.09 | — | S |
| Malathion | 121-75-5 | 8 | 466.50 | 0.0323 | 6.0138 | 0.85 | 139.10 | — | E |
| Mephosfolan | 950-10-7 | 1 | — | na | na | na | 0.14 | 1 | E |

Table 2. (Continued).

| Compound | CAS_RN | n | Median | Slope | Intercept | p | HD$_5$(50%) | #_EF | Status |
|---|---|---|---|---|---|---|---|---|---|
| Methacrifos | 62610-77-9 | 1 | — | na | na | na | 11.20 | 1 | E |
| Methamidophos | 10265-92-6 | 3 | 15.82 | na | na | na | 1.70 | 2 | E |
| Methidathion | 950-37-8 | 7 | 34.64 | na | na | na | 3.53 | 1 | E |
| Methiocarb | 2032-65-7 | 33 | 7.50 | 0.4708 | 0.0308 | 0.00 | 1.06 | — | E |
| Methocrotophos | 25601-84-7 | 2 | 3.12 | na | na | na | 0.25 | 1 | S |
| Methomyl | 16752-77-5 | 13 | 23.69 | 0.0852 | 2.7336 | 0.50 | 8.46 | — | E |
| Methyl parathion | 298-00-0 | 10 | 10.81 | 0.2021 | 1.3254 | 0.27 | 2.13 | — | E |
| Methyl trithion | 953-17-3 | 1 | — | na | na | na | 4.51 | 1 | S |
| Mevinphos | 7786-34-7 | 13 | 3.80 | 0.0254 | 0.9409 | 0.88 | 0.70 | — | E |
| Mexacarbate | 315-18-4 | 16 | 5.64 | -0.2586 | 3.2429 | 0.03 | 1.39 | — | S |
| Mobam | 10793-30-1 | 8 | 255.00 | 0.2555 | 4.1803 | 0.34 | 30.31 | — | S |
| Monocrotophos | 6923-22-4 | 23 | 2.51 | -0.0312 | 1.0218 | 0.79 | 0.42 | — | E |
| Naled | 300-76-5 | 7 | 64.90 | na | na | na | 1.72 | 1 | E |
| Omethoate | 1113-02-6 | 2 | 28.52 | na | na | na | 4.14 | 1 | E |
| Oxamyl | 23135-22-0 | 3 | 4.18 | na | na | na | 0.78 | 2 | E |
| Oxydemeton-methyl | 301-12-2 | 11 | 53.90 | -0.0452 | 4.1453 | 0.82 | 13.96 | — | E |
| Parathion | 56-38-2 | 19 | 5.62 | 0.0797 | 1.2228 | 0.76 | 0.40 | — | E |
| Phenthoate | 2597-03-7 | 2 | 259.00 | na | na | na | 23.17 | 1 | E |
| Phorate | 298-02-2 | 8 | 7.06 | 0.1817 | 0.4833 | 0.65 | 0.34 | — | E |
| Phosalone | 2310-17-0 | 1 | — | na | na | na | 106.27 | 1 | E |
| Phosfolan | 947-02-4 | 7 | 2.37 | 0.1034 | 0.8054 | 0.62 | 0.69 | — | S |
| Phosmet | 732-11-6 | 5 | 435.80 | 1.1854 | -1.5394 | 0.06 | 1.24 | — | E |
| Phosphamidon | 297-99-4 | 15 | 4.24 | 0.1528 | 0.7174 | 0.32 | 1.08 | — | E |
| Phoxim | 14816-18-3 | 12 | 32.21 | 0.6057 | 0.1756 | 0.01 | 1.71 | — | E |
| Pirimicarb | 23103-98-2 | 8 | 20.52 | 0.1118 | 2.4285 | 0.50 | 6.78 | — | E |
| Pirimiphos-ethyl | 23505-41-1 | 2 | 6.00 | na | na | na | 1.90 | 1 | E |

Table 2. (Continued).

| Compound | CAS_RN | $n$ | Median | Slope | Intercept | $p$ | $HD_5(50\%)$ | #_EF | Status |
|---|---|---|---|---|---|---|---|---|---|
| Pirimiphos-methyl | 29232-93-7 | 2 | 470.00 | na | na | na | 13.51 | 1 | E |
| Promecarb | 2631-37-0 | 4 | 13.50 | 1.3541 | -2.6021 | 0.15 | 0.94 | — | E |
| Propaphos | 7292-16-2 | 1 | — | na | na | na | 0.18 | 1 | E |
| Propetamphos | 31218-83-4 | 3 | 78.00 | na | na | na | 7.09 | 1 | E |
| Propoxur | 114-26-1 | 23 | 11.76 | 0.344 | 0.8344 | 0.02 | 1.31 | — | E |
| Prothiofos | 34643-46-4 | 1 | — | na | na | na | 13.65 | 1 | E |
| Prothoate (trimethoate) | 2275-18-5 | 2 | 55.95 | na | na | na | 5.52 | 1 | S |
| Pyridaphention | 119-12-0 | 1 | — | na | na | na | 7.94 | 1 | E |
| Pyrolan | 87-47-8 | 1 | — | na | na | na | 3.27 | 1 | S |
| Quinalphos | 13593-03-8 | 2 | 20.65 | na | na | na | 0.42 | 1 | E |
| Schradan | 152-16-9 | 3 | 19.00 | na | na | na | 2.02 | 1 | S |
| Sulfotep | 3689-24-5 | 2 | 150.00 | na | na | na | 50.63 | 1 | E |
| Sulprofos | 35400-43-2 | 5 | 47.00 | -0.0883 | 4.8854 | 0.89 | 6.85 | — | E |
| Tebupirimfos | 96182-53-5 | 1 | — | na | na | na | 2.36 | 1 | E |
| Temephos | 3383-96-8 | 14 | 65.60 | 0.3163 | 2.6217 | 0.10 | 8.68 | — | E |
| Terbufos | 13071-79-9 | 5 | 9.48 | 1.0277 | -1.5328 | 0.32 | 0.16 | — | E |
| Tetrachlorvinphos | 961-11-5 | 3 | 4750.00 | na | na | na | 25.32 | 1 | E |
| Thiocarboxime | 29118-87-4 | 5 | 12.00 | -0.4141 | 5.2618 | 0.08 | 5.04 | — | S |
| Thiodicarb | 59669-26-0 | 1 | — | na | na | na | 234.96 | 1 | E |
| Thiofanox | 39196-18-4 | 3 | 1.20 | na | na | na | 0.12 | 1 | E |
| Thiometon | 640-15-3 | 2 | 61.75 | na | na | na | 5.07 | 1 | E |
| Thionazin | 297-97-2 | 8 | 2.77 | -0.2254 | 2.2454 | 0.07 | 1.02 | 1 | E |
| Triazamate | 112143-82-5 | 1 | — | na | na | na | 0.93 | 1 | E |
| Triazophos | 24017-47-8 | 5 | 9.47 | 0.1991 | 1.0113 | 0.55 | 1.68 | — | E |
| Trichlorfon | 52-68-6 | 12 | 60.73 | 0.3189 | 2.4757 | 0.06 | 13.36 | — | E |
| Trichloronat | 327-98-0 | 10 | 18.50 | 0.4069 | 0.7989 | 0.30 | 0.73 | — | S |

Table 2. (Continued).

| Compound | CAS_RN | n | Median | Slope | Intercept | p | HD$_5$(50%) | #_EF | Status |
|---|---|---|---|---|---|---|---|---|---|
| Trimethacarb (Landrin™) | 12407-86-2 | 8 | 69.00 | -0.1382 | 5.0938 | 0.56 | 16.28 | — | E |
| Vamidothion | 2275-23-2 | 1 | — | na | na | na | 3.72 | 1 | E |
| Xylylcarb | 2425-10-7 | 1 | — | na | na | na | 6.20 | 1 | E |
| NUMBERED EXPERIMENTAL PRODUCTS: PROBABLY NEVER MARKETED | | | | | | | | | |
| 2-Propargyloxyphenyl N-methylcarbamate (HERCULES 9699) | 3279-46-7 | 2 | 45.00 | na | na | na | 11.39 | 1 | |
| 3-(2-Butyl)-phenyl N-methyl carbamate N-benzenesulfoate (RE-11775) | 25474-41-3 | 2 | 57.90 | na | na | na | 1.23 | 1 | |
| 3,4,5-Trimethylphenyl methylcarbamate (Landrin™ isomer 1) | 2686-99-9 | 1 | — | na | na | na | 4.50 | 1 | |
| 3,5-Diisopropylphenyl N-methylcarbamate (HRS 1422) | 330-64-3 | 2 | 55.00 | na | na | na | 2.53 | 1 | |
| 3,5 Xylyl N-methylcarbamate | 2655-14-3 | 1 | — | na | na | na | 19.61 | 1 | |
| 3-Propargyloxyphenyl N-methylcarbamate (HERCULES 8717) | 3692-90-8 | 2 | 40.48 | na | na | na | 3.80 | 1 | |

Table 2. (Continued).

| Compound | CAS_RN | n | Median | Slope | Intercept | p | HD$_5$(50%) | #_EF | Status |
|---|---|---|---|---|---|---|---|---|---|
| Diethyl 5-methylpyrazol-3yl phosphate (Pyrazoxon) | 108-34-9 | 1 | — | na | na | na | 3.38 | 1 | |
| m-Isopropylphenyl N-methylcarbamate (HERCULES 5727) | 64-00-6 | 2 | 11.31 | na | na | na | 1.42 | 1 | |
| O,O-Diethyl naphtalene-1,8-dicarboximido oxyphosphonothioate (BAY22408) | 2668-92-0 | 2 | 261.86 | na | na | na | 6.00 | 1 | |
| O,O-Dimethyl O-(3,5-dimethyl-4-methylthio-phenyl)phosphoro-thioate (Bayer 37342) | | 1 | — | na | na | na | 3.64 | 1 | |
| O,O-Dimethyl O-(4-methyl-mercaptophenyl) phosphate (GC-6506) | 3254-63-5 | 2 | 0.90 | na | na | na | 0.07 | 1 | |
| O-Ethyl S-p-tolyl ethyl phosphonodithioate (BAY 38156) | 333-43-7 | 2 | 3.34 | na | na | na | 0.43 | 1 | |
| Phenol, 3-(1-methyl-ethyl)-4-(methylthio)-methyl carbamate (ACD 7029) | | 7 | 13.30 | 0.0329 | 2.6034 | 0.94 | 1.20 | — | |

Table 2. (Continued).

| Compound | CAS_RN | n | Median | Slope | Intercept | p | HD$_5$(50%) | #_EF | Status |
|---|---|---|---|---|---|---|---|---|---|
| Phosphonamidothioic acid, p-ethyl-O-[3-methyl-4-(methylthio)-phenyl]ester (BAY HOL 0574) | | 9 | 5.62 | -0.2266 | 2.5104 | 0.41 | 0.51 | — | |
| Phosphoramidic acid, ethyl-,2,4-dichlorophenyl ester (DOWCO 161) | | 9 | 13.30 | 0.0151 | 2.8043 | 0.30 | 4.33 | — | |
| Phosphorothioic acid, O-(3-bromo-5,7-dimethyl-pyrazolo[1,5-a]-pirimidin-2-yl)O,O-diethyl ester (BAY 75546) | 7682-90-8 | 7 | 4.22 | 0.3415 | 0.035 | 0.00 | 0.91 | — | |
| **INSUFFICIENT PRODUCT DATA** | | | | | | | | | |
| Dimethylvinphos | 2274-67-1 | — | | | | | | | E |
| Isoxathion | 18854-01-8 | — | | | | | | | E |
| Mecarbam | 2595-54-2 | — | | | | | | | E |
| Metolcarb | 1129-41-5 | — | | | | | | | E |
| Profenofos | 41198-08-7 | — | | | | | | | E |
| Pyraclofos | 77458-01-6 | — | | | | | | | E |
| Dialifos | 10311-84-9 | 2 | 752.55 | | | | | | S |
| Formothion | 2540-82-1 | 2 | 238.85 | | | | | | E |
| Pirimiphos-methyl | 29232-93-7 | 1 | — | | | | | | E |

Table 3. Reference $LD_{50}$ values for pesticides that do not inhibit cholinesterase; the values are ordered by principal target pest and alphabetically by common chemical name.

| Chemical | CAS_RN | N | Median | Slope | Intercept | p | $HD_5(50\%)$ | #_EF | Status | Use |
|---|---|---|---|---|---|---|---|---|---|---|
| **CHEMICALS PRIMARILY ACTIVE AGAINST INSECTS AND OTHER INVERTEBRATES** | | | | | | | | | | |
| 2-(Octylthio)ethanol | 3547-33-9 | 2 | 2250.00 | — | — | — | 234.13 | 2 | E | Insecticide |
| Abamectin | 71751-41-2 | 2 | 1042.30 | — | — | — | 42.80 | 2 | E | Acaricide |
| Acequinocyl (AKD-2023) | 57960-19-7 | 2 | 2000.00 | — | — | — | 193.05 | 1 | E | Acaricide |
| Acetamiprid | 135410-20-7 | 1 | — | — | — | — | 20.91 | 1 | E | Insecticide |
| Acetates of Z/E 8-dodecenyl and Z 8-dodecenol (Z 8-dodecenyl) | 28079-04-1 | 1 | — | — | — | — | 232.29 | 1 | E | Insect pheromone |
| Acrinathrin | 101007-06-1 | 2 | 1625.00 | — | — | — | 156.09 | 2 | E | Acaricide insecticide |
| Aldrin | 309-00-2 | 12 | 19.83 | 0.44 | 0.80 | 0.15 | 1.15 | — | S | Insecticide |
| Allethrin | 584-79-2 | 1 | — | — | — | — | 192.68 | 1 | E | Insecticide |
| Alpha-Cypermethrin | 67375-30-8 | 1 | — | — | — | — | 9633.91 | 1 | E | Insecticide |
| Amitraz | 33089-61-1 | 1 | — | — | — | — | 41.83 | 1 | E | Insecticide acaricide |
| Azadirachtin | 992-20-1 | 1 | — | — | — | — | 261.32 | 1 | U | Miticide |
| Azocyclotin | 41083-11-8 | 1 | — | — | — | — | 17.42 | 1 | E | Acaricide |
| Bensultap | 17606-31-4 | 4 | 192.00 | 1.35 | -2.11 | 0.23 | 0.41 | — | E | Insecticide |
| Benzene Hexachloride | 608-73-1 | 1 | — | — | — | — | 12.54 | 1 | E | Insecticide |
| Benzyl benzoate | 120-51-4 | 1 | — | — | — | — | 232.29 | 1 | U | Miticide |
| Beta-Cyfluthrin | 68359-37-5 | 4 | — | — | — | — | — | | E | Insecticide |
| Bifenazate (D2341) | 149877-41-8 | 1 | — | — | — | — | 132.64 | 1 | E | Acaricide |
| Bifenthrin | 82657-04-3 | 2 | 1975.00 | — | — | — | 204.71 | 2 | E | Insecticide |

Also tabled are the CAS registration numbers, the number of species values available, the median $LD_{50}$, the slope, intercept and p value of the $LD_{50}*$weight regression, the calculated or estimated $HD_5(50\%)$ and, where used, the number of species from which an extrapolation factor was developed (see Table 1). The status follows the nomenclature of the *Pesticide Manual*, 11th Ed.; E, in use; S, superseded; U, unknown.

Table 3. (Continued).

| Chemical | CAS_RN | N | Median | Slope | Intercept | p | HD$_5$(50%) | #_EF | Status | Use |
|---|---|---|---|---|---|---|---|---|---|---|
| Binapacryl | 485-31-4 | 2 | 717.50 | — | — | — | — | — | S | Acaricide |
| Bioallethrin (Depallethrin) | 548-79-2 | 1 | — | — | — | — | 235.77 | 1 | E | Insecticide |
| Bioallethron S-cyclopentenyl isomer | 28434-00-6 | 3 | 5000.00 | — | — | — | 520.29 | 2 | E | Insecticide |
| Bioethanomethrin (RU-11-679) | 22431-62-5 | 1 | — | — | — | — | — | — | U | Insecticide |
| Bioresmethrin | 28434-01-7 | 1 | — | — | — | — | 506.33 | 1 | E | Insecticide |
| Bromopropylate | 18181-80-1 | 2 | 2255.00 | — | — | — | 193.05 | 1 | E | Acaricide |
| BT | 68038-71-1 | 1 | — | — | — | — | 481.70 | 1 | U | Insecticide |
| Buprofezin | 69327-76-0 | 3 | 2000.00 | — | — | — | 680.40 | 2 | E | Insecticide |
| Butoxypolypropylene glycol | 9003-13-8 | 1 | — | — | — | — | 261.32 | 1 | S | Insecticide |
| Calcium polysulfide | 1344-81-6 | 1 | — | — | — | — | 65.04 | 1 | E | Insecticide |
| Calcium tetrathiocarbamate | 81510-83-0 | 1 | — | — | — | — | 137.05 | 1 | U | Insecticide |
| CGA 50 439 | 61676-87-7 | 1 | — | — | — | — | 78.09 | 1 | E | Acaricide, ixodicide |
| Chlordane | 57-74-9 | 4 | 62.28 | 1.00 | -1.63 | 0.44 | 0.09 | — | E | Insecticide |
| Chlordecone | 143-50-0 | 2 | 220.33 | — | — | — | 26.42 | 1 | S | Insecticide, fungicide |
| Chlorfenapyr | 122453-73-0 | 2 | 8.30 | — | — | — | 0.56 | 1 | E | Insecticide, acaricide |
| Chlorfluazuron | 71422-67-8 | 1 | — | — | — | — | 241.81 | 1 | E | Insecticide |
| Chlorofenizon | 80-33-1 | 1 | — | — | — | — | 444.02 | 1 | S | Acaricide |
| Citronella oil | 8000-29-1 | 1 | — | — | — | — | 261.32 | 1 | E | Insecticide |
| Clofentezine | 74115-24-5 | 2 | 5250.00 | — | — | — | 493.59 | 2 | E | Acaricide |
| Codiemone | 33956-49-9 | 1 | — | — | — | — | 249.71 | 1 | E | Insect pheromone |

Table 3. (Continued).

| Chemical | CAS_RN | N | Median | Slope | Intercept | $p$ | HD$_5$(50%) | #_EF | Status | Use |
|---|---|---|---|---|---|---|---|---|---|---|
| Copper salts of fatty acids & rosin acids | 9007-39-0 | 1 | — | — | — | — | 209.06 | 1 | U | Insecticide |
| Cryolite | 15096-52-3 | 1 | — | — | — | — | 249.71 | 1 | E | Insecticide |
| Cycloprothrin | 63935-38-6 | 2 | 3500.00 | — | — | — | 482.63 | 1 | E | Insecticide |
| Cyfluthrin | 68359-37-5 | 4 | — | — | — | — | 485.44 | 1 | E | Insecticide |
| Cyhalothrin | 68085-85-8 | 1 | — | — | — | — | 481.70 | 1 | E | Insecticide |
| Cyhexatin | 13121-70-5 | 1 | — | — | — | — | 57.43 | 1 | E | Acaricide |
| Cypermethrin | 52315-07-8 | 3 | 10000.00 | — | — | — | 579.15 | 1 | E | Insecticide |
| Cyromazine | 66215-27-8 | 4 | 2061.50 | -0.13 | 8.25 | 0.63 | 604.60 | — | E | Insect growth regulator |
| DDT | 50-29-3 | 5 | 1334.00 | 0.36 | 5.12 | 0.37 | 122.90 | — | E | Insecticide |
| Deltamethrin | 52918-63-5 | 5 | 1000.00 | — | — | — | 97.09 | 1 | E | Insecticide |
| Diafenthiuron | 80060-09-9 | 2 | 1500.00 | — | — | — | 156.09 | 2 | E | Insecticide, acaricide |
| Dicofol | 115-32-2 | 3 | 680.99 | — | — | — | 72.37 | 1 | E | Acaricide |
| Dieldrin | 60-57-1 | 16 | 35.15 | 0.08 | 3.02 | 0.71 | 4.15 | — | S | Insecticide |
| Dienochlor | 2227-17-0 | 2 | 3087.95 | — | — | — | 243.97 | 2 | E | Insecticide |
| Diflubenzuron | 35367-38-5 | 3 | 5000.00 | — | — | — | 952.56 | 1 | E | Insecticide |
| Dinitro-O-cresol | 534-52-1 | 2 | 23.00 | — | — | — | 2.22 | 1 | U | Insecticide |
| Dipropyl isocinchomeronate | 136-45-8 | 1 | — | — | — | — | 156.79 | 1 | U | Insecticide |
| DUOMEEN T-E-9 (N-tallow-trimethylene diamines) | | 2 | 1019.00 | — | — | — | 37.83 | 1 | U | Mosquito control agent |
| Endosulfan | 115-29-7 | 6 | 52.42 | 0.34 | 2.23 | 0.17 | 9.53 | — | E | Insecticide |
| Endrin | 72-20-8 | 11 | 1.78 | 0.03 | 0.59 | 0.83 | 0.75 | — | S | Insecticide |
| Esfenvalerate | 66230-04-4 | 2 | 1478.51 | — | — | — | 131.24 | 2 | E | Insecticide |

Table 3. (Continued).

| Chemical | CAS_RN | N | Median | Slope | Intercept | p | HD$_5$(50%) | #_EF | Status | Use |
|---|---|---|---|---|---|---|---|---|---|---|
| Etofenprox | 80844-07-1 | 1 | — | — | — | — | 192.68 | 1 | E | Insecticide |
| Etoxazole | 153233-91-1 | 1 | — | — | — | — | 192.68 | 1 | E | Acaricide |
| Farnesol | 4602-84-0 | 2 | 2150.00 | — | — | — | 223.73 | 2 | E | Miticide |
| Fenazaquin | 120928-09-8 | 2 | 1873.50 | — | — | — | 194.51 | 2 | E | Acaricide |
| Fenbutatin oxide | 13356-08-6 | 5 | — | — | — | — | 291.52 | 1 | E | Acaricide |
| Fenothiocarb | 62850-32-2 | 2 | 1471.54 | — | — | — | 142.91 | 2 | E | Acaricide |
| Fenoxycarb | 79127-80-3 | 1 | — | — | — | — | 675.68 | 1 | E | Insecticide |
| Fenpyroximate | 111812-58-9 | 2 | 2000.00 | — | — | — | 208.12 | 2 | E | Acaricide |
| Fenvalerate | 51630-58-1 | 2 | 5447.36 | — | — | — | 321.77 | 2 | E | Insecticide |
| Fipronil | 120068-37-3 | 7 | 39.19 | -0.37 | 7.22 | 0.65 | 1.47 | — | E | Insecticide |
| Fluazuron | 86811-58-7 | 2 | 2000.00 | — | — | — | 208.12 | 2 | E | Ixodicide |
| Flubenzimine | 37893-02-0 | 1 | — | — | — | — | 431.67 | 1 | S | Acaricide |
| Flucycloxuron | 94050-52-9 | 1 | — | — | — | — | 192.68 | 1 | E | Acaricide, insecticide |
| Flucythrinate | 70124-77-5 | 2 | 2645.00 | — | — | — | 274.88 | 2 | E | Insecticide |
| Flufenoxuron | 101463-69-8 | 1 | — | — | — | — | 232.29 | 1 | E | Insecticide |
| Fluvalinate | 69409-94-5 | 1 | — | — | — | — | 291.52 | 1 | E | Insecticide |
| Gossyplure (Hexadeca-dienyl acetate) | 53042-79-8 | 1 | — | — | — | — | 963.39 | 1 | E | Insect pheromone |
| Halfenprox | 111872-58-3 | 1 | — | — | — | — | 218.82 | 1 | E | Acaricide |
| Heptachlor | 76-44-8 | 7 | 125.00 | -0.09 | 5.37 | 0.89 | 3.47 | — | E | Insecticide |
| Hexaflumuron | 86479-06-3 | 2 | 2000.00 | — | — | — | 208.12 | 2 | E | Insecticide |
| Hexythiazox | 78587-05-0 | 2 | 3620.27 | — | — | — | 482.63 | 1 | E | Acaricide |
| Hydramethylnon | 67485-29-4 | 2 | 2169.00 | — | — | — | 222.90 | 2 | E | Insecticide |
| Imazethabenz-methyl | 81405-85-8 | 2 | 2150.00 | — | — | — | 223.73 | 2 | E | Insecticide |
| Imidacloprid | 105827-78-9 | 7 | 35.36 | -0.06 | 3.90 | 0.79 | 8.43 | — | U | Insecticide |

Table 3. (Continued).

| Chemical | CAS_RN | N | Median | Slope | Intercept | p | HD$_5$(50%) | #_EF | Status | Use |
|---|---|---|---|---|---|---|---|---|---|---|
| Isobenzan | 297-78-9 | 7 | 3.16 | 0.41 | -0.90 | 0.12 | 0.43 | — | S | Insecticide |
| Isolan | 119-38-0 | 1 | — | — | — | — | 6.99 | 1 | S | Insecticide |
| Isomate-M | 28079-84-1 | 1 | — | — | — | — | 232.29 | 1 | U | Pheromone |
| Lambda-Cyhalothrin | 91465-08-6 | 1 | — | — | — | — | 428.14 | 1 | E | Insecticide |
| Limonene | 138-86-3 | 1 | — | — | — | — | 232.29 | 1 | U | Insecticide |
| Lindane | 58-89-9 | 11 | 90.83 | 0.42 | 2.86 | 0.14 | 10.50 | — | E | Insecticide |
| Lufenuron | 103055-07-8 | 2 | 2000.00 | — | — | — | 208.12 | 2 | E | Insecticide, acaricide |
| Methoprene | 40596-69-8 | 1 | — | — | — | — | 192.68 | 1 | E | Insect growth regulator |
| Methoxychlor | 72-43-5 | 4 | — | — | — | — | 291.52 | 1 | E | Insecticide |
| Methyl chloroform | 71-55-6 | 1 | — | — | — | — | 291.52 | 1 | U | Insecticide |
| Methyl nonyl ketone | 112-12-9 | 2 | 2250.00 | — | — | — | 234.13 | 2 | U | Insecticide |
| Muscalure | 27519-02-4 | 1 | — | — | — | — | 232.29 | 1 | E | Insect pheromone |
| N,N-Diethyl-M-Toluamide | 134-62-3 | 1 | — | — | — | — | 159.70 | 1 | U | Insecticide |
| Napthalene | 91-20-3 | 1 | — | — | — | — | 312.43 | 1 | S | Insecticide |
| Neurolidol | 7212-44-4 | 2 | 2150.00 | — | — | — | 223.73 | 2 | E | Miticide |
| Nicotine | 54-11-5 | 5 | 247.40 | 1.40 | -1.89 | 0.06 | 1.04 | — | E | Insecticide |
| Nicotine sulfate | 65-30-5 | 10 | 75.00 | 0.37 | 2.82 | 0.19 | 6.94 | — | U | Insecticide |
| Nitenpyram | 120738-89-8 | 2 | 1687.00 | — | — | — | 165.48 | 2 | E | Insecticide |
| Novaluron | 116714-46-6 | 1 | — | — | — | — | 192.68 | 1 | E | Insecticide |
| Paradichlorobenzene | 106-46-7 | 1 | — | — | — | — | 186.76 | 1 | U | Insecticide |
| Permethrin | 52645-53-1 | 5 | — | — | — | — | 3127.53 | 1 | E | Insecticide |
| Phenol | | 2 | 310.20 | — | — | — | — | — | U | Insecticide |
| Phenothrin [(1R)-trans-isomer] | 26002-80-2 | 1 | — | — | — | — | 291.52 | 1 | E | Insecticide |

Table 3. (Continued).

| Chemical | CAS_RN | N | Median | Slope | Intercept | p | HD$_5$(50%) | #_EF | Status | Use |
|---|---|---|---|---|---|---|---|---|---|---|
| Piperonyl butoxide | 51-03-6 | 1 | — | — | — | — | 261.32 | 1 | E | Insecticide |
| POE Isooctadecanol | 52292-17-8 | 1 | — | — | — | — | 192.68 | 1 | U | Insecticide |
| Polybutene | 9003-29-6 | 1 | — | — | — | — | 249.71 | 1 | U | Insecticide |
| Polychlorocamphanes | | 2 | 53.00 | — | — | — | — | — | U | Insecticide |
| Potassium salt of oleic acid | 143-18-0 | 1 | — | — | — | — | 246.38 | 1 | U | Insecticide |
| Potassium salts of fatty acids | 10124-65-9 | 1 | — | — | — | — | 291.52 | 1 | U | Insecticide |
| Prallethrin | 23031-36-9 | 2 | 1085.50 | — | — | — | 112.60 | 2 | E | Insecticide |
| Pymetrozine | 123312-89-0 | 2 | 2000.00 | — | — | — | 208.12 | 2 | E | Insecticide |
| Pyrethrin | 8003-34-7 | 1 | — | — | — | — | 963.39 | 1 | E | Insecticide |
| Pyridaben | 96489-71-3 | 3 | 2250.00 | — | — | — | 279.50 | 2 | E | Acaricide, insecticide |
| Pyrimidifen | 105779-78-0 | 1 | — | — | — | — | 42.87 | 1 | E | Acaricide, insecticide |
| Pyriproxyfen | 95737-68-1 | 2 | 2000.00 | — | — | — | 208.12 | 2 | E | Insecticide |
| Resmethrin | 10453-86-8 | 4 | — | — | — | — | 60.32 | 1 | E | Insecticide |
| Rotenone | 83-79-4 | 1 | — | — | — | — | 211.95 | 1 | E | Insecticide |
| Ryanodine | 15662-33-6 | 6 | 2.37 | 0.06 | 0.85 | 0.91 | 0.59 | — | S | Insecticide |
| SD-16898 | | 4 | 6.77 | -0.23 | 3.30 | 0.36 | 3.14 | — | U | Insecticide, acaricide |
| Silafluofen | 105024-66-9 | 2 | 2000.00 | — | — | — | 193.05 | 1 | E | Insecticide |
| Sodium cyanide | 143-33-9 | 6 | 8.99 | -0.08 | 2.64 | 0.77 | 2.13 | — | E | Insecticide, acaricide |
| Sulcofuron-sodium | 3567-25-7 | 2 | 1558.00 | — | — | — | 149.96 | 2 | E | Insecticide |
| Sulfluramid | 4151-50-2 | 1 | — | — | — | — | 16.96 | 1 | E | Insecticide |
| SZI-121 | 162320-67-4 | 1 | — | — | — | — | 193.05 | 1 | E | Acaricide |
| Tau-Fluvalinate | 102851-06-9 | 1 | — | — | — | — | 291.52 | 1 | E | Insecticide, acaricide |

Table 3. (Continued).

| Chemical | CAS_RN | N | Median | Slope | Intercept | p | HD$_5$(50%) | #_EF | Status | Use |
|---|---|---|---|---|---|---|---|---|---|---|
| TDE | 72-54-8 | 1 | — | — | — | — | 41.02 | 1 | S | Insecticide, mosquito larvicide and adulticide |
| Tebufenozide | 112410-23-8 | 1 | — | — | — | — | 249.71 | 1 | E | Insecticide |
| Tebufenpyrad | 119168-77-3 | 2 | 2000.00 | — | — | — | 208.12 | 2 | E | Acaricide |
| Teflubenzuron | 83121-18-0 | 2 | 2125.00 | — | — | — | 220.74 | 2 | E | Insecticide |
| Tefluthrin | 79538-32-2 | 3 | 734.00 | — | — | — | 178.63 | 2 | E | Insecticide |
| TEPA | 545-55-1 | 7 | 29.90 | -0.24 | 4.71 | 0.26 | 4.45 | — | U | Insecticide, chemo-sterilant |
| Tetradifon | 116-29-0 | 4 | — | — | — | — | 580.72 | 1 | E | Acaricide |
| Tetramethrin | 7696-12-0 | 1 | — | — | — | — | 276.01 | 1 | E | Insecticide |
| Thiocyclam | 31895-21-3 | 1 | — | — | — | — | 333.01 | 1 | E | Insecticide |
| Tobacco dust | 8037-19-2 | 1 | — | — | — | — | 249.71 | 1 | U | Insecticide |
| Toxaphene | 8001-35-2 | 11 | 70.70 | -0.32 | 6.26 | 0.21 | 10.40 | — | S | Insecticide |
| Tralomethrin | 66841-25-6 | 1 | — | — | — | — | 291.52 | 1 | E | Insecticide |
| Transfluthrin | 118712-89-3 | 1 | — | — | — | — | — | — | E | Insecticide |
| Tridec-4-en-1-yl acetate | 65954-19-0 | 2 | 2060.66 | — | — | — | 214.34 | 2 | E | Insecticide |
| Triflumuron | 64628-44-0 | 2 | 2780.50 | — | — | — | 208.05 | 2 | E | Insecticide |
| Z-11-Hexadecanol | 53939-28-9 | 1 | — | — | — | — | 232.29 | 1 | U | Insecticide |
| ZXI 8901 | 160791-64-0 | 1 | — | — | — | — | 32.53 | 1 | E | Insecticide |

CHEMICALS PRIMARILY ACTIVE AGAINST BACTERIA AND ALGAE OR APPLIED AS FUMIGANTS

| Chemical | CAS_RN | N | Median | Slope | Intercept | p | HD$_5$(50%) | #_EF | Status | Use |
|---|---|---|---|---|---|---|---|---|---|---|
| 1,2-Benzenedicarbox-aldehyde | 643-79-8 | 1 | — | — | — | — | 71.10 | 1 | U | Bactericide |

Table 3. (Continued).

| Chemical | CAS_RN | N | Median | Slope | Intercept | p | HD$_5$(50%) | #_EF | Status | Use |
|---|---|---|---|---|---|---|---|---|---|---|
| 1,3-dibromo-5,5-dimethylhydantoin (DBDMH) | 77-48-5 | 1 | — | — | — | — | 291.52 | 1 | U | Bactericide |
| 2-(Hydroxymethyl)amino)ethanol | 34375-28-5 | 1 | — | — | — | — | 202.44 | 1 | U | Bactericide |
| 2-Benzyl-4-chlorophenol | 120-32-1 | 1 | — | — | — | — | 291.52 | 1 | U | Bactericide |
| 4,4-Dimethyloxazolidine | 51200-87-4 | 2 | 905.00 | — | — | — | 91.84 | 2 | U | Bactericide |
| 4-Chloro-3,5-xylenol | 88-04-0 | 1 | — | — | — | — | 276.01 | 1 | U | Bactericide |
| Acrolein | 107-02-8 | 2 | 14.06 | — | — | — | 1.37 | 2 | E | Herbicide, fungicide, microbicide |
| ADBAC | 68424-85-1 | 2 | 810.04 | — | — | — | 49.61 | 2 | U | Bactericide |
| Alkyl amine hydrochloride | 91745-52-7 | 1 | — | — | — | — | 114.87 | 1 | U | Bactericide |
| Alkyl amino-3-aminopropane | 61791-64-8 | 1 | — | — | — | — | 8.56 | 1 | U | Bactericide |
| Alkyl trimethyl ammonium chloride | 61789-18-2 | 1 | — | — | — | — | 62.95 | 1 | U | Bactericide |
| Azadioxabicyclooctane | 59720-42-2 | 1 | — | — | — | — | 241.81 | 1 | U | Bactericide |
| BCDMH | 16079-88-2 | 1 | — | — | — | — | 187.84 | 1 | U | Bactericide |
| Benzisothiazolin-3-one | 2634-33-5 | 1 | — | — | — | — | 71.66 | 1 | U | Bactericide |
| BHAP (Bromohydroxya-cetophenone) | 2491-38-5 | 1 | — | — | — | — | 76.89 | 1 | U | Bactericide |
| Bioban P-1487 | 2224-44-4 | 1 | — | — | — | — | 96.34 | 1 | U | Bactericide |
| Bis(bromoacetoxy)-2-butene | 20679-58-7 | 1 | — | — | — | — | 18.88 | 1 | U | Bactericide |

Table 3. (Continued).

| Chemical | CAS_RN | N | Median | Slope | Intercept | p | HD$_5$(50%) | #_EF | Status | Use |
|---|---|---|---|---|---|---|---|---|---|---|
| Bis(trichloromethyl) sulfone | 3064-70-8 | 1 | — | — | — | — | 216.76 | 1 | U | Bactericide |
| Boric acid | 10043-35-3 | 1 | — | — | — | — | 291.52 | 1 | U | Pesticide |
| Bromonitrostyrene | 7166-19-0 | 1 | — | — | — | — | 48.17 | 1 | U | Bactericide |
| Bronopol | 52-51-7 | 1 | — | — | — | — | 49.04 | 1 | E | Bactericide |
| Busan 77 | 31512-74-0 | 1 | — | — | — | — | 47.88 | 1 | U | Bactericide |
| Calcium hypochlorite | 7778-54-3 | 2 | 803.50 | — | — | — | 41.32 | 2 | U | Algicide |
| Chlorhexidine diacetate | 56-95-1 | 1 | — | — | — | — | 233.80 | 1 | U | Bactericide |
| Chromic acid | 7738-94-5 | 1 | — | — | — | — | 19.05 | 1 | U | Preservative |
| Copper sulfate (basic) | 1332-14-5 | 1 | — | — | — | — | 47.12 | 1 | U | Antifoulant |
| Copper sulfate pentahydrate | 7758-99-8 | 1 | — | — | — | — | 43.89 | 1 | E | Algicide |
| Copper triethanolamine | 82027-59-6 | 1 | — | — | — | — | 192.68 | 1 | U | Bactericide |
| Cosan 145 | 97553-90-7 | 1 | — | — | — | — | 156.79 | 1 | U | Preservative |
| Cuprous thiocyanate | 1111-67-7 | 1 | — | — | — | — | 232.29 | 1 | E | Bactericide |
| Dazomet | 533-74-4 | 3 | 383.50 | — | — | — | 53.33 | 1 | E | Fumigant |
| DBNPA | 10222-01-2 | 2 | 220.43 | — | — | — | 22.91 | 2 | U | Bactericide |
| DCDIC | 138-93-2 | 1 | — | — | — | — | 309.34 | 1 | U | Bactericide |
| DCDMH (1,3-Dichloro-5,5-dimethylhydantoin) | 118-52-5 | 1 | — | — | — | — | 291.52 | 1 | U | Bactericide |
| DDAC | 7173-51-5 | 1 | — | — | — | — | 12.62 | 1 | U | Bactericide |
| Dichloropropene/methylisothiocyanate | 542-75-6, 556-61-6 | 1 | — | — | — | — | 36.80 | 1 | | Fumigant, fungicide, nematicide |
| Diiodomethyl p-tolyl sulfone | 20018-09-1 | 1 | — | — | — | — | 291.52 | 1 | U | Preservative |
| Dimethoxane | 828-00-2 | 1 | — | — | — | — | 184.09 | 1 | U | Bactericide |

Table 3. (Continued).

| Chemical | CAS_RN | N | Median | Slope | Intercept | p | HD$_5$(50%) | #_EF | Status | Use |
|---|---|---|---|---|---|---|---|---|---|---|
| Dioctyl dimethyl ammonium chloride | 5538-94-5 | 1 | — | — | — | — | 20.35 | 1 | U | Algicide |
| Dithio-3-one,4,5-dichloro | 1192-52-5 | 1 | — | — | — | — | 28.69 | 1 | U | Bactericide |
| Dowicil | 4080-31-3 | 2 | 1795.00 | — | — | — | 183.10 | 2 | U | Bactericide |
| DTEA | 36362-09-1 | 1 | — | — | — | — | 216.76 | 1 | U | Bactericide |
| Erioglaucine/tartrazine | 3844-45-9 | 2 | 2250.00 | — | — | — | 234.13 | 2 | U | Algicide |
| Formaldehyde | 50-00-0 | 1 | — | — | — | — | 91.75 | 1 | E | Fungicide, bactericide |
| Glutaraldehyde | 111-30-8 | 1 | — | — | — | — | 167.89 | 1 | U | Bactericide |
| Grotan | 4719-04-4 | 1 | — | — | — | — | 176.54 | 1 | U | Bactericide |
| Guanidine (dodine free base) | 112-65-2 | 2 | 1391.87 | — | — | — | 141.62 | 2 | E | Bactericide |
| Hydroxypropyl methane thiosulfonate | 30388-01-3 | 1 | — | — | — | — | 45.66 | 1 | U | Bactericide |
| Iodine | 7553-56-2 | 1 | — | — | — | — | 232.29 | 1 | U | Bactericide |
| Iodine complex | 11096-42-7 | 1 | — | — | — | — | 197.44 | 1 | U | Bactericide |
| Irgarol | 28159-98-0 | 1 | — | — | — | — | 261.32 | 1 | U | Bactericide |
| Isobardac | 138698-36-9 | 1 | — | — | — | — | 3.95 | 1 | U | Bactericide |
| Isocyanuric acid | 108-80-5 | 1 | — | — | — | — | 184.49 | 1 | U | Bactericide |
| Lithium hypochlorite | 13840-33-0 | 1 | — | — | — | — | 54.62 | 1 | U | Bactericide |
| Metaldehyde | 108-62-3 | 3 | 196.00 | — | — | — | 16.93 | 1 | E | Molluscicide |
| Methyl bromide | 74-83-9 | 1 | — | — | — | — | 8.48 | 1 | E | Soil sterilant, fumigant |
| Methyl isocyanate | 556-61-6 | 1 | — | — | — | — | 13.10 | 1 | E | Fungicide, nematicide, fumigant |

Table 3. (Continued).

| Chemical | CAS_RN | N | Median | Slope | Intercept | p | HD$_5$(50%) | #_EF | Status | Use |
|---|---|---|---|---|---|---|---|---|---|---|
| Methylisothiazolinone (Acticide 14) | 26172-55-4 | 1 | — | — | — | — | 7.91 | 1 | U | Bactericide |
| Metronidazole | 443-48-1 | 1 | — | — | — | — | 481.70 | 1 | U | Bactericide |
| Mineral oil (including parafin oil) | 8012-95-1 | 1 | — | — | — | — | 261.32 | 1 | U | Bactericide |
| MTI | 82633-79-2 | 1 | — | — | — | — | 17.65 | 1 | U | Preservative |
| Nemagon | 96-12-8 | 2 | 111.40 | — | — | — | 16.58 | 1 | S | Nematicide |
| Nicosamide-olamine | 1420-04-8 | 3 | — | — | — | — | 232.29 | 1 | E | Molluscicide |
| Nitrapyrin | 1929-82-4 | 1 | — | — | — | — | 260.89 | 1 | E | Bactericide, nitrification inhibitor |
| OBPA (DID 47) | 58-36-6 | 2 | 5050.00 | — | — | — | 963.39 | 1 | U | Bactericide |
| Octhilinone | 26530-20-1 | 1 | — | — | — | — | 55.50 | 1 | E | Bactericide |
| Oxazolidine E | 7747-35-5 | 1 | — | — | — | — | 116.14 | 1 | U | Bactericide |
| Parachlorometacresol | 59-50-7 | 1 | — | — | — | — | 178.86 | 1 | U | Bactericide |
| PHMB | 32289-58-0 | 1 | — | — | — | — | 241.81 | 1 | E | Bactericide |
| PNMDC/DCDMC | 137-41-7 | 1 | — | — | — | — | 114.86 | 1 | U | Bactericide |
| Potassium azide | 12136-44-6 | 3 | 20.80 | — | — | — | 1.60 | 1 | U | MISC |
| Potassium bromide | 7758-02-3 | 1 | — | — | — | — | 290.36 | 1 | U | Bactericide |
| Potassium dimethylthio-carbamate | 128-03-0 | 1 | — | — | — | — | 145.76 | 1 | U | Bactericide |
| Pyrazole | 85264-33-1 | 1 | — | — | — | — | 88.15 | 1 | U | Bactericide |
| SDDC | 128-04-1 | 1 | — | — | — | — | 115.10 | 1 | U | Bactericide |
| Silver | 7440-22-4 | 1 | — | — | — | — | 261.32 | 1 | U | Bactericide |
| Sodium 2-mercaptobenzothiolate | 2492-26-4 | 1 | — | — | — | — | 249.71 | 1 | U | Bactericide |
| Sodium 2-phenylphenate | 132-27-4 | 1 | — | — | — | — | 116.14 | 1 | U | Bactericide |

Table 3. (Continued).

| Chemical | CAS_RN | N | Median | Slope | Intercept | p | $HD_5(50\%)$ | #_EF | Status | Use |
|---|---|---|---|---|---|---|---|---|---|---|
| Sodium bromide | 7647-15-6 | 2 | 2200.00 | — | — | — | 228.87 | 2 | U | Bactericide |
| Sodium chlorite | 7758-19-2 | 2 | 792.86 | — | — | — | 79.64 | 2 | U | Bactericide |
| Sodium dichloro-S-tri-azinetrione | 2893-78-9 | 2 | 1954.68 | — | — | — | 203.36 | 2 | U | Bactericide |
| Sodium dichloroisocyan-uratedihydrate | 51580-86-0 | 1 | — | — | — | — | 206.27 | 1 | U | Bactericide |
| Sodium dodecylbenzene-sulfonate | 25155-30-0 | 1 | — | — | — | — | 157.49 | 1 | U | Bactericide |
| Sodium hypochlorite | 7681-52-9 | 2 | 2710.14 | — | — | — | 272.16 | 2 | U | Bactericide |
| Sodium omadine | 15922-78-8 | 2 | 194.49 | — | — | — | 17.20 | 2 | U | Bactericide |
| Streptomycin | 3810-74-0 | 1 | — | — | — | — | 232.29 | 1 | E | Bactericide |
| TBT methacrylate | 26354-18-7 | 1 | — | — | — | — | 192.68 | 1 | U | Antifoulant |
| TCMTB | 21564-17-0 | 1 | — | — | — | — | 76.75 | 1 | U | Bactericide |
| Tetraglycine hydroperi-odide | 7097-60-1 | 1 | — | — | — | — | 29.04 | 1 | U | Bactericide |
| THPS | 55566-30-8 | 1 | — | — | — | — | 29.58 | 1 | U | Bactericide |
| Tributyltin methacrylate | 2155-70-6 | 1 | — | — | — | — | 81.07 | 1 | U | Bactericide |
| Trichloro-s-triazinetrione | 87-90-1 | 2 | 1480.87 | — | — | — | 146.48 | 2 | U | Bactericide |
| Trichloromelamine | 7673-09-8 | 1 | — | — | — | — | 246.38 | 1 | U | Bactericide |
| Triclosan | 3380-34-5 | 2 | 1487.50 | — | — | — | 138.59 | 2 | U | Bactericide |
| Triethylhexahydro-s-triazine | 7779-27-3 | 2 | 472.52 | — | — | — | 47.49 | 2 | U | Preservative |
| Trimethoxysilyl quats | 27668-52-6 | 1 | — | — | — | — | 153.18 | 1 | U | Bactericide |
| Zinc naphthenate | 12001-85-3 | 1 | — | — | — | — | 261.32 | 1 | U | Preservative |
| Zinc oxide | 1314-13-2 | 1 | — | — | — | — | 65.74 | 1 | U | Preservative |

Table 3. (Continued).

### CHEMICALS PRIMARILY ACTIVE AGAINST FUNGI

| Chemical | CAS_RN | N | Median | Slope | Intercept | p | HD$_5$(50%) | #_EF | Status | Use |
|---|---|---|---|---|---|---|---|---|---|---|
| 2-Phenylphenol | 132-27-4 | 1 | — | — | — | — | 116.14 | 1 | E | Fungicide |
| 3-Iodo-2-propynyl butyl-carbamate | 55406-53-6 | 1 | — | — | — | — | 99.00 | 1 | U | Fungicide |
| Anilazine | 101-05-3 | 5 | 2000.00 | 0.19 | 6.73 | 0.21 | 614.90 | — | S | Fungicide |
| Azaconazole | 60207-31-0 | 2 | 731.00 | — | — | — | 48.26 | 1 | E | Fungicide |
| Azoxystrobin | 131860-33-8 | 1 | — | — | — | — | 232.29 | 1 | E | Fungicide |
| Barium metaborate | 13701-59-2 | 1 | — | — | — | — | 145.64 | 1 | U | Fungicide |
| Benalaxyl | 71626-11-4 | 2 | 6850.00 | — | — | — | 819.31 | 2 | E | Fungicide |
| Benomyl | 17804-35-2 | 3 | 100.00 | — | — | — | 25.32 | 1 | E | Fungicide |
| Bis tributyltin oxide | 56-35-9 | 1 | — | — | — | — | 68.80 | 1 | S | Fungicide |
| Bitertanol | 55179-31-2 | 3 | 2000.00 | — | — | — | 420.74 | 1 | E | Fungicide |
| Bromuconazol | 116255-48-2 | 2 | 2150.00 | — | — | — | 223.73 | 2 | E | Fungicide |
| Bupirimate | 41483-43-6 | 2 | 3873.50 | — | — | — | 482.63 | 1 | E | Fungicide |
| Captafol | 2425-06-1 | 1 | — | — | — | — | 291.52 | 1 | E | Fungicide |
| Captan | 133-06-2 | 3 | 100.00 | — | — | — | 25.32 | 1 | E | Fungicide |
| Carbendazim | 10605-21-7 | 2 | 4681.99 | — | — | — | 491.11 | 2 | E | Fungicide |
| Carboxin | 5234-68-4 | 5 | 2000.00 | 1.64 | -1.95 | 0.04 | 3.44 | — | S | Fungicide |
| Ceresan M | 517-16-8 | 1 | — | — | — | — | 33.48 | 1 | S | Fungicide |
| Chinomethionat (oxythio-quinox) | 2439-01-2 | 3 | 500.00 | — | — | — | 126.58 | 1 | E | Acaricide, fungicide |
| Chloroneb | 2675-77-6 | 1 | — | — | — | — | 481.70 | 1 | S | Fungicide |
| Chlorothalonil | 1897-45-6 | 1 | — | — | — | — | 193.05 | 1 | E | Fungicide |
| Chlozolinate | 84332-86-5 | 3 | 4500.00 | — | — | — | 790.55 | 2 | E | Fungicide |
| Copper hydroxide | 20427-59-2 | 3 | 3400.00 | — | — | — | 219.11 | 2 | E | Fungicide |
| Copper napthenate | 1338-02-9 | 1 | — | — | — | — | 261.32 | 1 | E | Fungicide |

Table 3. (Continued).

| Chemical | CAS_RN | N | Median | Slope | Intercept | p | HD₅(50%) | #_EF | Status | Use |
|---|---|---|---|---|---|---|---|---|---|---|
| Copper oxychloride | 1332-40-7 | 1 | — | — | — | — | 45.95 | 1 | E | Fungicide |
| Cuprobame | 7076-63-3 | 1 | — | — | — | — | — | 1 | S | Fungicide |
| Cuprous oxide | 1317-39-1 | 2 | 406.00 | — | — | — | 33.16 | 1 | E | Fungicide |
| Cycloheximide | 66-81-9 | 3 | 9.38 | — | — | — | 1.85 | 2 | S | Fungicide |
| Cymoxanil | 57966-95-7 | 2 | 2250.00 | — | — | — | 234.13 | 2 | E | Fungicide |
| Cyproconazole | 113096-99-4 | 1 | — | — | — | — | 14.22 | 2 | E | Fungicide |
| Cyprodinil | 121552-61-2 | 2 | 2000.00 | — | — | — | 208.12 | 2 | E | Fungicide |
| Debacarb/carbendazim | 62732-91-6 | 1 | — | — | — | — | 96.34 | 1 | E | Fungicide |
| Dichlofluanid | 1085-98-9 | 2 | 5000.00 | — | — | — | 482.63 | 1 | E | Fungicide |
| Dichlone | 117-80-6 | 1 | — | — | — | — | 192.68 | 1 | S | Fungicide |
| Dichloropropene | 542-75-6 | 1 | — | — | — | — | 17.65 | 1 | E | Fungicide |
| Dichlobutrazol | 75736-33-3 | 1 | — | — | — | — | 928.81 | 1 | E | Fungicide |
| Dicloran | 99-30-9 | 3 | 900.00 | — | — | — | 139.61 | 2 | E | Fungicide |
| Diethofencarb | 87130-20-9 | 2 | 2250.00 | — | — | — | 234.13 | 2 | E | Fungicide |
| Difenoconazol | 119446-68-3 | 1 | — | — | — | — | 207.13 | 1 | E | Fungicide |
| Diflumetorim | 130339-07-0 | 2 | 1430.00 | — | — | — | 85.04 | 1 | E | Fungicide |
| Dimethirimol | 5221-53-4 | 1 | — | — | — | — | 202.53 | 1 | E | Fungicide |
| Dimethomorph | 110488-70-5 | 2 | 2000.00 | — | — | — | 208.12 | 2 | E | Fungicide |
| Diniconazole | 83657-24-3 | 2 | 1745.10 | — | — | — | 179.64 | 2 | E | Fungicide |
| Dinobuton | 973-21-7 | 1 | — | — | — | — | 7.59 | 1 | E | Acaricide, fungicide |
| Dinocap | 39300-45-3 | 1 | — | — | — | — | 249.71 | 1 | E | Fungicide, acaricide |
| Dithianon | 3347-22-6 | 6 | 294.50 | 0.50 | 2.66 | 0.51 | 5.29 | — | E | Fungicide |
| DMDM hydantoin | 6440-58-0 | 1 | — | — | — | — | 141.62 | 1 | U | Fungicide |
| Dodine (doguadine) | 2439-10-3 | 1 | — | — | — | — | 110.02 | 1 | E | Fungicide |
| Edifenphos | 17109-49-8 | 1 | — | — | — | — | 75.14 | 1 | E | Fungicide |

Table 3. (Continued).

| Chemical | CAS_RN | N | Median | Slope | Intercept | p | HD$_5$(50%) | #_EF | Status | Use |
|---|---|---|---|---|---|---|---|---|---|---|
| Epoxiconazol | 106325-08-0 | 1 | — | — | — | — | 232.29 | 1 | E | Fungicide |
| Ethirimol | 23947-60-6 | 1 | — | — | — | — | 202.53 | 1 | E | Fungicide |
| Etridiazole | 2593-15-9 | 1 | — | — | — | — | 65.04 | 1 | E | Fungicide |
| Fenaminosulf | 140-56-7 | 3 | 17.80 | — | — | — | 4.52 | 1 | S | Fungicide |
| Fenarimol | 60168-88-9 | 2 | 632.46 | — | — | — | 65.81 | 2 | E | Fungicide |
| Fenbuconazol | 114369-43-6 | 2 | 2174.72 | — | — | — | 270.13 | 2 | E | Fungicide |
| Fenfuram | 24691-80-3 | 1 | — | — | — | — | 0.69 | 1 | E | Fungicide |
| Fenpiclonil | 74738-17-3 | 2 | 1305.00 | — | — | — | 52.13 | 2 | E | Fungicide |
| Fenpropidin | 67306-00-7 | 1 | — | — | — | — | 39.21 | 1 | E | Fungicide |
| Fenpropimorph | 67306-03-0 | 3 | 3949.68 | — | — | — | 419.73 | 1 | E | Fungicide |
| Fentin acetate | 900-95-8 | 4 | 107.00 | 0.38 | 2.90 | 0.05 | 38.25 | — | E | Fungicide |
| Fentin hydroxide | 76-87-9 | 3 | 76.07 | — | — | — | 7.39 | 2 | E | Fungicide |
| Ferimzone | 89269-64-7 | 1 | — | — | — | — | 94.14 | 1 | E | Fungicide |
| Fluazinam | 79622-59-6 | 2 | 2986.00 | — | — | — | 284.34 | 2 | E | Fungicide |
| Fludioxonil | 131341-86-1 | 2 | 2000.00 | — | — | — | 208.12 | 2 | E | Fungicide |
| Fluquinconazole | 136426-54-5 | 2 | 2000.00 | — | — | — | 208.12 | 2 | E | Fungicide |
| Flusilazole | 85509-19-9 | 1 | — | — | — | — | 153.18 | 1 | E | Fungicide |
| Flusulfamide | 106917-52-6 | 1 | — | — | — | — | 7.67 | 1 | E | Fungicide |
| Flutolanil | 66332-96-5 | 2 | 2000.00 | — | — | — | 208.12 | 2 | E | Fungicide |
| Flutriafol | 76674-21-0 | 2 | 2808.00 | — | — | — | 481.70 | 1 | E | Fungicide |
| Folpet | 133-07-3 | 2 | 1405.78 | — | — | — | 118.28 | 2 | E | Fungicide |
| Fosetyl-aluminium | 39148-24-8 | 2 | 6498.50 | — | — | — | 785.42 | 2 | E | Fungicide |
| Fuberidazole | 3878-19-1 | 2 | 424.28 | — | — | — | 47.78 | 1 | E | Fungicide |
| Furalaxyl | 57646-30-7 | 1 | — | — | — | — | 579.15 | 1 | E | Fungicide |
| Guazatine | 13516-27-3 | 3 | 216.00 | — | — | — | 41.99 | 1 | E | Fungicide |
| Guazatine (triacetate) | 57520-17-9 | 3 | 120.00 | — | — | — | 32.65 | 1 | U | Fungicide |

Table 3. (Continued).

| Chemical | CAS_RN | N | Median | Slope | Intercept | p | HD$_5$(50%) | #_EF | Status | Use |
|---|---|---|---|---|---|---|---|---|---|---|
| Hexaconazole | 79983-71-4 | 1 | — | — | — | — | 391.14 | 1 | E | Fungicide |
| Hymexazol | 10004-44-1 | 3 | 1479.00 | — | — | — | 175.77 | 2 | E | Fungicide |
| Imazalil | 35554-44-0 | 3 | 2000.00 | — | — | — | 49.23 | 1 | E | Fungicide |
| Imibenconazole | 86598-92-7 | 2 | 2250.00 | — | — | — | 234.13 | 2 | E | Fungicide |
| Iminoctadine triacetate | 39202-40-9 | 1 | — | — | — | — | 94.89 | 1 | E | Fungicide |
| Iprodione | 36734-19-7 | 1 | — | — | — | — | 158.40 | 1 | E | Fungicide |
| Isoprothiolane | 50512-35-1 | 1 | — | — | — | — | 428.29 | 1 | E | Fungicide |
| Kasugamycin | 6980-18-3 | 1 | — | — | — | — | 386.10 | 1 | E | Fungicide, bactericide |
| Lignasan BLP | 52316-55-9 | 1 | — | — | — | — | 447.01 | 1 | U | Fungicide |
| Mancozeb | 8018-01-7 | 3 | 6861.49 | — | — | — | 710.95 | 1 | E | Fungicide |
| Maneb | 12427-38-2 | 4 | — | — | — | — | 345.34 | 1 | E | Fungicide |
| Mepanipyrim | 110235-47-7 | 2 | 2250.00 | — | — | — | 234.13 | 2 | E | Fungicide |
| Mepronil | 55814-41-0 | 2 | 2000.00 | — | — | — | 208.12 | 2 | E | Fungicide |
| Mercuric chloride | 7487-94-7 | 1 | — | — | — | — | 4.10 | 1 | E | Fungicide |
| Metalaxyl | 57837-19-1 | 1 | — | — | — | — | 89.09 | 1 | E | Fungicide |
| Metalaxyl-M | 70630-17-0 | 1 | — | — | — | — | 137.03 | 1 | E | Fungicide |
| Metconazole | 125116-23-6 | 1 | — | — | — | — | 91.75 | 1 | E | Fungicide |
| Methylmercury dicyan-diamide | 502-39-6 | 2 | 35.00 | — | — | — | 5.11 | 1 | S | Fungicide |
| Metiram | 9006-42-2 | 1 | — | — | — | — | 249.71 | 1 | E | Fungicide |
| Myclobutanil | 88671-89-0 | 1 | — | — | — | — | 59.23 | 1 | E | Fungicide |
| Nabam | 142-59-6 | 3 | 2120.00 | — | — | — | 204.63 | 1 | E | Fungicide |
| Nuarimol | 63284-71-9 | 1 | — | — | — | — | 73.46 | 1 | E | Fungicide |
| Ofurace | 58810-48-3 | 1 | — | — | — | — | — | | E | Fungicide |
| Oxadixyl | 77732-09-3 | 3 | 2000.00 | — | — | — | 149.07 | 2 | E | Fungicide |

Table 3. (Continued).

| Chemical | CAS_RN | N | Median | Slope | Intercept | p | HD$_5$(50%) | #_EF | Status | Use |
|---|---|---|---|---|---|---|---|---|---|---|
| Oxine-copper | 10380-28-6 | 3 | 618.00 | — | — | — | 115.69 | 2 | E | Fungicide |
| Oxytetracycline | 79-57-2 | 1 | — | — | — | — | 232.29 | 1 | S | Fungicide |
| Paranitrophenol | 100-02-7 | 1 | — | — | — | — | 67.02 | 1 | U | Fungicide |
| Pefurazoate | 101903-30-4 | 2 | 3300.00 | — | — | — | 229.73 | 1 | E | Fungicide |
| Penconazole | 66246-88-6 | 3 | 2424.00 | — | — | — | 193.41 | 2 | E | Fungicide |
| Pencycuron | 66063-05-6 | 3 | 2000.00 | — | — | — | 277.77 | 2 | E | Fungicide |
| Pentachlorophenol (PCP) | 87-86-5 | 3 | 504.00 | — | — | — | 50.79 | 2 | U | Insecticide, fungicide, herbicide |
| Phenylmercuric acetate (PMA) | 62-38-4 | 4 | 145.86 | 2.19 | -9.43 | 0.16 | 0.01 | — | E | Fungicide |
| Prochloraz | 67747-09-5 | 3 | 707.00 | — | — | — | 74.20 | 2 | E | Fungicide |
| Prochloraz-manganese | | 1 | — | — | — | — | 202.31 | 1 | U | Fungicide |
| Procymidone | 32809-16-8 | 2 | 5300.00 | — | — | — | 637.07 | 1 | E | Fungicide |
| Propamocarb | 24579-73-5 | 2 | 2910.00 | — | — | — | 321.72 | 1 | E | Fungicide |
| Propiconazole | 60207-90-1 | 4 | 2667.50 | 0.33 | 5.69 | 0.29 | 296.80 | — | E | Fungicide |
| Propineb | 12071-83-9 | 2 | 3000.00 | — | — | — | 482.63 | 1 | E | Fungicide |
| Pyrazophos | 13457-18-6 | 2 | 295.53 | — | — | — | 26.53 | 1 | E | Fungicide |
| Pyrifenox | 88283-41-4 | 2 | 2000.00 | — | — | — | 208.12 | 2 | E | Fungicide |
| Pyrimethanil | 53112-28-0 | 2 | 2000.00 | — | — | — | 208.12 | 2 | E | Fungicide |
| Pyrithione | 1121-30-8 | 2 | 126.93 | — | — | — | 25.32 | 1 | S | Fungicide |
| Quintozene | 82-68-8 | 1 | — | — | — | — | 255.45 | 1 | E | Fungicide |
| Sec-butylamine | 13952-84-6 | 1 | — | — | — | — | 12.66 | 1 | E | Fungicide |
| Silicate methoxyethyl mercury | | 3 | 18.00 | — | — | — | 1.91 | 1 | U | Fungicide |
| Sodium tetrathioperoxo-carbonate (GY-81) | 7345-69-9 | 1 | — | — | — | — | 137.05 | 1 | E | Fungicide, insecticide, nematicide |

Table 3. (Continued).

| Chemical | CAS_RN | N | Median | Slope | Intercept | p | HD₅(50%) | #_EF | Status | Use |
|---|---|---|---|---|---|---|---|---|---|---|
| Tebuconazole | 107534-96-3 | 4 | 1494.00 | 0.32 | 5.97 | 0.19 | 347.30 | — | E | Fungicide |
| Terrazole | 2593-15-9 | 2 | 1100.00 | — | — | — | 99.72 | 2 | E | Fungicide |
| Thiabendazole | 148-79-8 | 1 | — | — | — | — | 261.32 | 1 | E | Fungicide |
| Thiophanate-methyl | 23564-05-8 | 1 | — | — | — | — | 482.63 | 1 | E | Fungicide |
| Thiram | 137-26-8 | 7 | 500.00 | 0.55 | 3.31 | 0.06 | 36.81 | — | E | Fungicide |
| Tolclofos-methyl | 57018-04-9 | 3 | 5000.00 | — | — | — | 485.06 | 2 | E | Fungicide |
| Tolylfluanid | 731-27-1 | 4 | — | — | — | — | 482.63 | 1 | E | Fungicide |
| Trans-1,2-bis(n-propylsulfonyl)ethene (CHE 1843) | 1113-14-0 | 2 | 1560.00 | — | — | — | 192.68 | 1 | U | Fungicide |
| Triadimefon | 43121-43-3 | 4 | — | — | — | — | 385.36 | — | E | Fungicide |
| Triadimenol | 55219-65-3 | 3 | 2000.00 | — | — | — | 555.54 | 2 | E | Fungicide |
| Triazoxide | 72459-58-6 | 3 | 108.43 | — | — | — | 10.47 | 1 | E | Fungicide |
| Trichogramma harzianum | | 2 | 2000.00 | — | — | — | 208.12 | 2 | E | Fungicide |
| Tricyclazole | 41814-78-2 | 2 | 100.00 | — | — | — | 10.41 | 2 | E | Fungicide |
| Tridemorph | 81412-43-3 | 2 | 1694.00 | — | — | — | 173.37 | 2 | E | Fungicide |
| Triflumizole | 68694-11-1 | 1 | — | — | — | — | 291.52 | 1 | E | Fungicide |
| Triforine | 26644-46-2 | 5 | — | — | — | — | 776.70 | 1 | E | Fungicide |
| Triticonazole | 131983-72-7 | 6 | — | — | — | — | 232.29 | 1 | E | Fungicide |
| Vinclozolin | 50471-44-8 | 1 | — | — | — | — | 291.52 | 1 | E | Fungicide |
| Zinc borate | 12447-61-9 | 1 | — | — | — | — | 261.32 | 1 | U | Fungicide |
| Zineb | 12122-67-7 | 1 | — | — | — | — | 212.54 | 1 | E | Fungicide |
| Zirame | 137-30-4 | 4 | 88.73 | -0.25 | 5.94 | 0.36 | 29.45 | — | E | Fungicide |
| CHEMICALS PRIMARILY ACTIVE AGAINST PLANTS | | | | | | | | | | |
| 2,3,6-TBA | 50-31-7 | 2 | 1353.55 | — | — | — | 75.14 | 1 | E | Herbicide |
| 2,4,5-T | 93-76-5 | 1 | — | — | — | — | 56.52 | 1 | S | Herbicide |

Table 3. (Continued).

| Chemical | CAS_RN | N | Median | Slope | Intercept | p | HD$_5$(50%) | #_EF | Status | Use |
|---|---|---|---|---|---|---|---|---|---|---|
| 2,4-D | 94-75-7 | 6 | 531.74 | 0.13 | 5.57 | 0.68 | 132.90 | — | E | Herbicide |
| 2,4-D Butotyl | 1929-73-3 | 1 | — | — | — | — | 232.29 | 1 | E | Herbicide |
| 2,4-D Dimethylammonium | 2008-39-1 | 1 | — | — | — | — | 58.07 | 1 | E | Herbicide |
| 2,4-D diolamine | 5742-19-8 | 1 | — | — | — | — | 69.11 | 1 | U | Herbicide |
| 2,4-D Isooctyl ester | 1928-43-4 | 1 | — | — | — | — | 63.87 | 1 | E | Herbicide |
| 2,4-D sodium | 2702-72-9 | 1 | — | — | — | — | 195.09 | 1 | E | Herbicide |
| 2,4-D Tri-isopropylamine salt | 32341-80-3 | 1 | — | — | — | — | 47.04 | 1 | U | Herbicide |
| 2,4-DB | 94-82-6 | 1 | — | — | — | — | 178.40 | 1 | E | Herbicide |
| 2,4-DP-p DMA salt | 104786-87-0 | 1 | — | — | — | — | 32.40 | 1 | U | Herbicide |
| 2,4-DP Dimethylamine salt | 53404-32-3 | 1 | — | — | — | — | 32.40 | 1 | U | Herbicide |
| 6-Benzylaminopurine (N6-Benzuladenine) | 1214-39-7 | 1 | — | — | — | — | 185.71 | 1 | E | Plant growth regulator |
| Acetochlor | 34256-82-1 | 2 | 1133.37 | — | — | — | 96.27 | 2 | E | Herbicide |
| Acibenzolat (CGA 245704) | 135158-54-2 | 1 | — | — | — | — | 232.29 | 1 | E | Plant activator |
| Acifluorfen-sodium | 62476-59-9 | 2 | 1573.00 | — | — | — | 99.64 | 2 | E | Herbicide |
| Aclonifen | 74070-46-5 | 1 | — | — | — | — | 1447.88 | 1 | E | Herbicide |
| Alachlor | 15972-60-8 | 1 | — | — | — | — | 330.42 | 1 | E | Herbicide |
| Alloxydim-sodium | 55634-91-8 | 1 | — | — | — | — | 286.20 | 1 | E | Herbicide |
| Ametryn | 834-12-8 | 2 | 3445.00 | — | — | — | 336.22 | 2 | E | Herbicide |
| Amidosulfuron | 120923-37-7 | 3 | 2000.00 | — | — | — | 248.45 | 2 | E | Herbicide |
| Amitrole | 61-82-5 | 2 | 3575.00 | — | — | — | 407.29 | 2 | E | Herbicide |
| Ammonium sulfamate | 7773-06-0 | 1 | — | — | — | — | 289.58 | 1 | E | Herbicide |
| Ancymidol | 12771-68-5 | 1 | — | — | — | — | 25.32 | 1 | E | Plant growth regulator |

Table 3. (Continued).

| Chemical | CAS_RN | N | Median | Slope | Intercept | p | HD$_5$(50%) | #_EF | Status | Use |
|---|---|---|---|---|---|---|---|---|---|---|
| Anilofos | 64249-01-0 | 1 | — | — | — | — | 270.60 | 1 | E | Herbicide |
| Arsenic acid | 7778-39-4 | 1 | — | — | — | — | 4.23 | 1 | U | Herbicide |
| Asulam | 3337-71-1 | 4 | — | — | — | — | 1668.64 | 1 | E | Herbicide |
| Asulam sodium | 2302-17-2 | 4 | — | — | — | — | 425.80 | 1 | E | Herbicide |
| Atrazine | 1919-24-9 | 2 | 7118.50 | — | — | — | 408.98 | 1 | E | Herbicide |
| Azafenidin | 68049-83-2 | 2 | 2250.00 | — | — | — | 234.13 | 2 | E | Herbicide |
| Azimsulfuron | 120162-55-2 | 2 | 2250.00 | — | — | — | 234.13 | 2 | E | Herbicide |
| Benazolin-ethyl | 25059-80-7 | 2 | 4500.00 | — | — | — | 441.48 | 2 | E | Herbicide |
| Benazolin | 3813-05-6 | 1 | — | — | — | — | 984.56 | 1 | E | Herbicide |
| Benfluralin | 1861-40-1 | 2 | 2000.00 | — | — | — | 208.12 | 2 | E | Herbicide |
| Benfuresate | 68505-69-1 | 2 | 21136.00 | — | — | — | 1869.35 | 2 | E | Herbicide |
| Benoxacor | 98730-04-2 | 2 | 2075.00 | — | — | — | 215.78 | 2 | E | Herbicide safener |
| Bensulfuron-methyl | 83055-99-6 | 1 | — | — | — | — | 207.13 | 1 | E | Herbicide |
| Bensulide | 741-58-2 | 1 | — | — | — | — | 160.98 | 1 | E | Herbicide |
| Bentazon | 50723-80-3 | 4 | 2029.00 | 0.91 | 2.41 | 0.14 | 32.40 | — | E | Herbicide |
| Bifenox | 42576-02-3 | 3 | 5000.00 | — | — | — | 407.29 | 2 | E | Herbicide |
| Bilanafos | 35597-43-4 | 1 | — | — | — | — | 253.16 | 1 | E | Herbicide |
| Bispyribac-sodium | 125401-75-4 | 1 | — | — | — | — | 261.32 | 1 | E | Herbicide |
| Bromacil | 314-40-9 | 1 | — | — | — | — | 261.32 | 1 | E | Herbicide |
| Bromoxynil | 1689-84-5 | 2 | 208.50 | — | — | — | 21.68 | 2 | E | Herbicide |
| Bromoxynil-potassium | 2961-68-4 | 1 | — | — | — | — | 5.31 | 1 | E | Herbicide |
| Bromoxynil (butyrate) | 3861-41-4 | 2 | 825.00 | — | — | — | 74.79 | 2 | U | Herbicide |
| Bromoxynil heptanoate | 56634-95-8 | 1 | — | — | — | — | 41.70 | 1 | U | Herbicide |
| Bromoxynil octanoate | 1689-99-2 | 3 | 170.00 | — | — | — | 16.20 | 2 | E | Herbicide |
| Bromoxynil Phenol | 1689-84-5 | 3 | 200.00 | — | — | — | 21.68 | 2 | U | Herbicide |
| Butachlor | 23184-66-9 | 1 | — | — | — | — | 447.01 | 1 | E | Herbicide |

Table 3. (Continued).

| Chemical | CAS_RN | N | Median | Slope | Intercept | p | HD$_5$(50%) | #_EF | Status | Use |
|---|---|---|---|---|---|---|---|---|---|---|
| Butralin | 33629-47-9 | 2 | 3625.00 | — | — | — | 416.66 | 2 | E | Herbicide |
| Butroxydim | 138164-12-2 | 2 | 1610.50 | — | — | — | 162.61 | 2 | E | Herbicide |
| Cafenstrole | 125306-83-4 | 1 | — | — | — | — | 481.70 | 1 | E | Herbicide |
| Capric acid/pelargonic acid | 143-07-7 | 1 | — | — | — | — | 261.32 | 1 | U | Herbicide |
| Carbetamide | 16118-49-3 | 2 | 2000.00 | — | — | — | 212.54 | 1 | E | Herbicide |
| Chlomethoxyfen | 32861-85-1 | 1 | — | — | — | — | — | | E | Herbicide |
| Chloramben | 133-90-4 | 1 | — | — | — | — | 159.40 | 1 | E | Herbicide |
| Chloramben-ammonium | 1076-46-6 | 1 | 2000.00 | — | — | — | 192.68 | 1 | U | Herbicide |
| Chloramben-sodium | 1954-81-0 | 2 | 2000.00 | — | — | — | 208.12 | 2 | U | Herbicide |
| Chlorflurenol | 2536-31-4 | 2 | — | — | — | — | 1437.86 | 1 | S | Growth regulator |
| Chloridazon | 1698-60-8 | 2 | 3000.00 | — | — | — | 351.36 | 2 | E | Herbicide |
| Chlorimuron-ethyl | 90982-32-4 | 1 | — | — | — | — | 241.81 | 1 | E | Herbicide |
| Chlormequat | 7003-89-6 | 2 | 408.00 | — | — | — | 53.57 | 1 | E | Plant growth regulator |
| Chlormequat chloride | 999-81-5 | 5 | 555.00 | -0.43 | 9.04 | 0.16 | 155.00 | — | E | Herbicide |
| Chloroprop-sodium | 53404-22-1 | 2 | 2334.00 | — | — | — | 218.55 | 2 | U | Growth regulator |
| Chloroxuron | 1982-47-4 | 2 | 4519.00 | — | — | — | 293.24 | 1 | S | Herbicide |
| Chlorpropham | 101-21-3 | 2 | 2000.00 | — | — | — | 193.05 | 1 | E | Herbicide |
| Chlorsulfuron | 64902-72-3 | 1 | — | — | — | — | 481.70 | 1 | E | Herbicide |
| Chlorthal-dimethyl | 1861-32-1 | 1 | — | — | — | — | 261.32 | 1 | E | Herbicide |
| Chlorthiamid | 1918-13-4 | 1 | — | — | — | — | 25.32 | 1 | E | Herbicide |
| Cinmethylin | 87818-31-3 | 1 | — | — | — | — | 249.71 | 1 | E | Herbicide |
| Cinosulfuron | 94593-91-6 | 2 | 2000.00 | — | — | — | 193.05 | 1 | E | Herbicide |
| Clethodim | 99129-21-2 | 1 | — | — | — | — | 232.29 | 1 | E | Herbicide |
| Clodinafop-propargyl | 105512-06-9 | 2 | 1727.50 | — | — | — | 177.51 | 2 | E | Herbicide |

Table 3. (Continued).

| Chemical | CAS_RN | N | Median | Slope | Intercept | p | HD5(50%) | #_EF | Status | Use |
|---|---|---|---|---|---|---|---|---|---|---|
| Clofencet | 129025-54-3 | 2 | 1707.00 | — | — | — | 174.99 | 2 | E | Plant growth regulator |
| Clomazone | 81777-89-1 | 2 | 2510.00 | — | — | — | 261.19 | 2 | E | Herbicide |
| Clopyralid | 1702-17-6 | 2 | 1855.86 | — | — | — | 192.53 | 2 | E | Herbicide |
| Cloquintocet-mexyl | 99607-70-2 | 2 | 2000.00 | — | — | — | 208.12 | 2 | E | Herbicide safender |
| Cyanazine | 21725-46-2 | 3 | 445.00 | — | — | — | 52.41 | 2 | E | Herbicide |
| Cyclanilide | 113136-77-9 | 2 | 215.50 | — | — | — | 22.42 | 2 | E | Plant growth regulator |
| Cycloate | 1134-23-2 | 4 | — | — | — | — | 249.71 | 1 | E | Herbicide |
| Cycloxydim | 101205-02-1 | 2 | 2000.00 | — | — | — | 208.12 | 2 | E | Herbicide |
| Cytokinin | 525-79-1 | 1 | — | — | — | — | 291.52 | 1 | U | Growth regulator |
| Daimuron | 42609-52-9 | 1 | — | — | — | — | 232.29 | 1 | E | Herbicide |
| Dalapon-sodium | 127-20-8 | 1 | — | — | — | — | 286.58 | 1 | E | Herbicide |
| Daminozide | 1596-84-5 | 1 | — | — | — | — | 311.28 | 1 | E | Plant growth regulator |
| Desmedipham | 13684-56-5 | 1 | — | — | — | — | 258.67 | 1 | E | Herbicide |
| Dicamba | 1918-00-9 | 2 | 936.55 | — | — | — | 62.26 | 2 | E | Herbicide |
| Dicamba-aluminum | | 1 | — | — | — | — | 241.81 | 1 | U | Herbicide |
| Dicamba-diglycoamine | 104040-79-1 | 1 | — | — | — | — | 44.97 | 1 | U | Herbicide |
| Dicamba-dimethylammonium | 2300-66-5 | 1 | — | — | — | — | 241.81 | 1 | E | Herbicide |
| Dicamba-potassium | 10007-85-9 | 1 | — | — | — | — | 31.94 | 1 | E | Herbicide |
| Dichlobenil | 1194-65-6 | 4 | 499.61 | 0.54 | 2.75 | 0.39 | 9.33 | — | E | Herbicide |
| Dichlorprop-P | 15165-67-0 | 1 | — | — | — | — | 41.06 | 1 | E | Herbicide |
| Diclofop-methyl | 51338-27-3 | 2 | 7971.85 | — | — | — | 788.26 | 2 | E | Herbicide |
| Diflufenican | 83164-33-4 | 2 | 3075.00 | — | — | — | 305.16 | 2 | E | Herbicide |

Table 3. (Continued).

| Chemical | CAS_RN | N | Median | Slope | Intercept | p | HD$_5$(50%) | #_EF | Status | Use |
|---|---|---|---|---|---|---|---|---|---|---|
| Diflufenzopyr (BAS 654 00 H) | 109293-98-3 | 1 | — | — | — | — | 541.43 | 1 | E | Herbicide |
| Dikegulac-sodium | 52508-35-7 | 1 | — | — | — | — | 374.86 | 1 | E | Herbicide |
| Dimepiperate | 61432-55-1 | 2 | 3500.00 | — | — | — | 193.05 | 1 | E | Herbicide |
| Dimethenamid (SAN 582) | 87674-68-8 | 1 | — | — | — | — | 221.60 | 1 | E | Herbicide |
| Dimethipin | 55290-64-7 | 1 | — | — | — | — | 84.78 | 1 | E | Herbicide, plant growth regulator |
| Dinitramine | 29091-05-2 | 2 | 5600.00 | — | 0.31 | — | 360.47 | 2 | E | Herbicide |
| Dinoseb | 88-85-7 | 5 | 19.90 | 0.42 | | 0.01 | 3.18 | — | S | Herbicide |
| Dinoseb acetate | 2813-95-8 | 2 | 16.66 | — | — | — | 2.29 | 1 | S | Herbicide |
| Dinoterb | 1420-07-1 | 2 | 25.00 | — | — | — | — | — | E | Herbicide |
| Diquat (dibromide) | 85-00-7 | 3 | 184.87 | — | — | — | 17.81 | 1 | E | Herbicide |
| Disodium methane-arsonate | 144-21-8 | 1 | — | — | — | — | 72.82 | 1 | E | Herbicide |
| Dithiopyr | 97886-45-8 | 1 | — | — | — | — | 261.32 | 1 | E | Herbicide |
| Diuron | 330-54-1 | 3 | 2000.00 | — | — | — | 193.04 | 2 | E | Herbicide |
| DMPA | 299-85-4 | 1 | — | — | — | — | 25.32 | 1 | S | Herbicide |
| DNOC | 534-52-1 | 5 | 21.15 | 0.15 | 1.97 | 0.36 | 6.67 | — | E | Herbicide |
| Endothall | 145-73-3 | 2 | 326.72 | — | — | — | 29.20 | 2 | E | Herbicide, algicide, plant growth regulator |
| Endothall (dimethylal-kylamine) | 66330-88-9 | 1 | — | — | — | — | 37.48 | 1 | U | Herbicide, algicide, plant growth regulator |

Table 3. (Continued).

| Chemical | CAS_RN | N | Median | Slope | Intercept | p | HD$_5$(50%) | #_EF | Status | Use |
|---|---|---|---|---|---|---|---|---|---|---|
| Endothall (dipotassium salt) | 2164-07-0 | 1 | — | — | — | — | 31.60 | 1 | U | Herbicide, algicide, plant growth regulator |
| EPTC | 759-94-4 | 3 | 1000.00 | — | — | — | 25.32 | 1 | E | Herbicide |
| Esprocarb | 85785-20-2 | 1 | — | — | — | — | 193.05 | 1 | E | Herbicide |
| Ethalfluralin | 55283-68-6 | 1 | — | — | — | — | 232.29 | 1 | E | Herbicide |
| Ethametsulfuron-methyl | 97780-06-8 | 1 | — | — | — | — | 261.32 | 1 | E | Herbicide |
| Ethephon | 16672-87-0 | 4 | 1924.00 | 0.21 | 6.24 | 0.48 | 372.20 | — | E | Herbicide |
| Ethidimuron | 30043-49-3 | 4 | 513.21 | 0.78 | 2.69 | 0.03 | 59.64 | — | S | Herbicide |
| Ethofumesate | 26255-79-6 | 3 | 3464.10 | — | — | — | 472.79 | 2 | E | Herbicide |
| Fenchlorazole | 103112-36-3 | 2 | 1333.84 | — | — | — | 25.84 | 1 | S | Herbicide |
| Fenclorim | 3740-92-9 | 2 | 2750.00 | — | — | — | 48.26 | 1 | E | Herbicide safener |
| Fenoprop | 93-72-1 | 2 | 9320.00 | — | — | — | 1091.75 | 1 | S | Herbicide |
| Fenoxaprop | 95617-09-7 | 3 | 2510.00 | — | — | — | 440.07 | 2 | S | Herbicide |
| Fenoxaprop-ethyl | 66441-23-4 | 3 | 2510.00 | — | — | — | 440.07 | 2 | S | Herbicide |
| Fenoxaprop-P | 113158-40-0 | 3 | — | — | — | — | 232.29 | 2 | E | Herbicide |
| Fenoxaprop-P-ethyl | 71283-80-2 | 1 | — | — | — | — | 232.29 | 1 | E | Herbicide |
| Fenridazone-sodium | 83588-43-6 | 1 | — | — | — | — | 495.70 | 1 | U | Herbicide |
| Fenuron | 101-42-8 | 4 | — | — | — | — | 232.29 | 1 | E | Herbicide |
| Ferric sulfate (see Ferrous sulfate) | 10028-22-5 | 1 | — | — | — | — | 261.32 | 1 | U | Herbicide |
| Ferrous sulfate heptahydrate | 7782-63-0 | 1 | — | — | — | — | 261.32 | 1 | E | Herbicide |
| Ferrous sulfate monohydrate | 17375-41-6 | 1 | — | — | — | — | 249.71 | 1 | U | Herbicide |
| Flamprop-M-methyl | 63729-98-6 | 1 | — | — | — | — | 447.88 | 1 | E | Herbicide |

Table 3. (Continued).

| Chemical | CAS_RN | N | Median | Slope | Intercept | $p$ | $HD_5(50\%)$ | #_EF | Status | Use |
|---|---|---|---|---|---|---|---|---|---|---|
| Flamprop-methyl | 52756-25-9 | 3 | 1000.00 | — | — | — | 106.27 | 1 | S | Herbicide |
| Flamprop isopropyl | 52756-22-6 | 2 | 1500.00 | — | — | — | 106.27 | 1 | S | Herbicide |
| Flazasulfuron | 104040-78-0 | 1 | — | — | — | — | 193.05 | 1 | E | Herbicide |
| Fluazifop-butyl | 69806-50-4 | 1 | — | — | — | — | 746.09 | 1 | E | Herbicide |
| Fluazifop-P-butyl | 79241-46-6 | 1 | — | — | — | — | 339.88 | 1 | E | Herbicide |
| Flumetralin | 62924-70-3 | 2 | 2150.00 | — | — | — | 223.73 | 2 | E | Growth regulator |
| Flumetsulam | 98967-40-9 | 1 | — | — | — | — | 261.32 | 1 | E | Herbicide |
| Flumiclorac-pentyl | 87546-18-7 | 1 | — | — | — | — | 261.32 | 1 | E | Herbicide |
| Fluometuron | 2164-17-2 | 1 | — | — | — | — | 192.68 | 1 | E | Herbicide |
| Fluoroglycofen | 77501-60-1 | 1 | — | — | — | — | 214.07 | 1 | E | Herbicide |
| Fluoroglycofen-ethyl | 77501-90-7 | 1 | — | — | — | — | 367.02 | 1 | E | Herbicide |
| Fluoxypyr-meptyl | 81406-37-3 | 2 | 2000.00 | — | — | — | 208.12 | 2 | E | Herbicide |
| Flupoxam | 119126-15-7 | 2 | 2250.00 | — | — | — | 234.13 | 2 | E | Herbicide |
| Flupyrsulfuron-methyl-sodium | 144740-54-5 | 1 | — | — | — | — | 216.76 | 1 | E | Herbicide |
| Flurazole | 72850-64-7 | 1 | — | — | — | — | 291.52 | 1 | E | Herbicide safender |
| Fluridone | 59756-60-4 | 1 | — | — | — | — | 232.29 | 1 | E | Herbicide |
| Flurochloridone | 61213-25-0 | 1 | — | — | — | — | 249.71 | 1 | E | Herbicide |
| Fluroxypyr | 69377-81-7 | 2 | 2000.00 | — | — | — | 208.12 | 2 | E | Herbicide |
| Flurprimidol | 56425-91-3 | 1 | — | — | — | — | 232.29 | 1 | E | Plant growth regulator |
| Fluxofenim | 88485-37-4 | 1 | — | — | — | — | 232.29 | 1 | E | Herbicide safender |
| Fomesafen | 72178-02-0 | 1 | — | — | — | — | 481.70 | 1 | E | Herbicide |
| Forchlorfenuron | 68157-60-8 | 1 | — | — | — | — | 261.32 | 1 | E | Plant growth regulator |

Table 3. (Continued).

| Chemical | CAS_RN | N | Median | Slope | Intercept | p | HD$_5$(50%) | #_EF | Status | Use |
|---|---|---|---|---|---|---|---|---|---|---|
| Fosamine ammonium | 25954-13-6 | 2 | 4791.29 | — | — | — | 498.10 | 2 | E | Herbicide |
| Furilazole | 121776-33-8 | 1 | — | — | — | — | 23.23 | 1 | E | Herbicide safender |
| Gibberellic acid | 77-06-5 | 1 | — | — | — | — | 261.32 | 1 | E | Growth regulator |
| Glufosinate-ammonium | 51276-47-2 | 4 | — | — | — | — | 232.29 | 1 | E | Herbicide |
| Glyphosate | 1071-83-6 | 1 | — | — | — | — | 232.29 | 1 | E | Herbicide |
| Glyphosate-trimesium (sulfosate) | 81591-81-3 | 2 | 1487.50 | — | — | — | 144.33 | 2 | E | Herbicide |
| Halosulfuron-methyl | 100784-20-1 | 1 | — | — | — | — | 261.32 | 1 | E | Herbicide |
| Haloxyfop | 69806-34-4 | 1 | — | — | — | — | 207.13 | 1 | E | Herbicide |
| Haloxyfop-P-methyl | 72619-32-0 | 1 | — | — | — | — | 134.61 | 1 | E | Herbicide |
| Haloxyfop ethoxyethyl (etotyl) | 87237-48-7 | 1 | — | — | — | — | 207.13 | 1 | E | Herbicide |
| Hexaflurate | 17029-22-0 | 3 | 193.00 | — | — | — | 18.59 | 1 | S | Herbicide |
| Hexazinone | 51235-04-2 | 1 | — | — | — | — | 261.44 | 1 | E | Herbicide |
| Hydrogen cyanamide | 420-04-2 | 1 | — | — | — | — | 28.74 | 1 | E | Herbicide, plant growth regulator |
| ICIS-0748 | 81052-29-1 | 2 | 2150.00 | — | — | — | 223.73 | 2 | E | Growth regulator |
| Imazamethabenz | 100728-84-5 | 2 | 2150.00 | — | — | — | 223.73 | 2 | E | Herbicide |
| Imazamethabenz-methyl | 81405-85-8 | 2 | 2150.00 | — | — | — | 223.73 | 2 | E | Herbicide |
| Imazamox | 114311-32-9 | 1 | — | — | — | — | 214.40 | 1 | E | Herbicide |
| Imazapic (AC 263,222) | 104098-49-9 | 2 | 2150.00 | — | — | — | 223.73 | 2 | E | Herbicide |
| Imazapyr | 81334-34-1 | 2 | 2150.00 | — | — | — | 223.73 | 2 | E | Herbicide |
| Imazaquine | 81335-37-7 | 2 | 2150.00 | — | — | — | 223.73 | 2 | E | Herbicide |
| Imazethapyr | 81335-77-5 | 2 | 2150.00 | — | — | — | 223.73 | 2 | E | Herbicide |
| Imazosulfuron | 122548-33-8 | 2 | 2250.00 | — | — | — | 234.13 | 2 | E | Herbicide |

Table 3. (Continued).

| Chemical | CAS_RN | N | Median | Slope | Intercept | p | HD$_5$(50%) | #_EF | Status | Use |
|---|---|---|---|---|---|---|---|---|---|---|
| Indole-3-butyric acid | 133-32-4 | 1 | — | — | — | — | 249.71 | 1 | E | Plant growth regulator |
| Ioxynil | 1689-83-4 | 4 | 68.50 | 0.10 | 3.68 | 0.54 | 33.28 | — | E | Herbicide |
| Ioxynil octanoate | 3861-47-0 | 4 | 385.42 | 0.29 | 4.19 | 0.64 | 19.41 | — | E | Herbicide |
| Isopropalin | 33820-53-0 | 1 | — | — | — | — | 232.29 | 1 | S | Herbicide |
| Isoproturon | 34123-59-6 | 3 | 4543.04 | — | — | — | 313.40 | 2 | E | Herbicide |
| Isouron | 55861-78-4 | 1 | — | — | — | — | 232.29 | 1 | E | Herbicide |
| Isoxaben | 82558-50-7 | 1 | — | — | — | — | 232.29 | 1 | E | Herbicide |
| Isoxaflutole | 141112-29-0 | 1 | — | — | — | — | 207.13 | 1 | E | Herbicide |
| Karbutilate | 4849-32-5 | 1 | — | — | — | — | 579.15 | 1 | S | Herbicide |
| L-Lactic acid | 50-21-5 | 1 | — | — | — | — | 261.32 | 1 | U | Growth regulator |
| Lactofen | 77501-63-4 | 1 | — | — | — | — | 291.52 | 1 | E | Herbicide |
| Lenacil | 2164-08-1 | 1 | — | — | — | — | — | — | E | Herbicide |
| Linuron | 330-55-2 | 3 | 505.00 | — | — | — | 65.07 | 2 | E | Herbicide |
| Maleic hydrazide potassium salt | 51542-52-0 | 1 | — | — | — | — | 216.76 | 1 | E | Plant growth regulator |
| MCPA | 94-74-6 | 2 | 377.00 | — | — | — | 39.23 | 2 | E | Herbicide |
| MCPA-thioethyl | 25319-90-8 | 1 | — | — | — | — | 289.58 | 1 | E | Herbicide |
| MCPA dimethylamine salt | 2039-46-5 | 1 | — | — | — | — | 55.54 | 1 | U | Herbicide |
| MCPB-sodium | 6062-26-6 | 1 | — | — | — | — | 32.75 | 1 | E | Herbicide |
| MCPP Isooctyl ester | 28473-03-2 | 1 | — | — | — | — | 261.32 | 1 | U | Herbicide |
| Mecoprop | 7085-19-0 | 1 | — | — | — | — | 82.11 | 1 | E | Herbicide |
| Mecoprop-P | 16484-77-8 | 2 | 2361.00 | — | — | — | 151.19 | 2 | E | Herbicide |
| Mecoprop dimethylamine salt | 32351-70-5 | 1 | — | — | — | — | 69.92 | 1 | U | Herbicide |

Table 3. (Continued).

| Chemical | CAS_RN | N | Median | Slope | Intercept | p | HD$_5$(50%) | #_EF | Status | Use |
|---|---|---|---|---|---|---|---|---|---|---|
| Mefenpyr-diethyl | 135590-91-9 | 1 | — | — | — | — | 193.05 | 1 | E | Herbicide safener |
| Mefluidide | 53780-36-2 | 1 | — | — | — | — | 353.05 | 1 | E | Plant growth regulator, herbicide |
| Mepiquat chloride | 24307-26-4 | 1 | — | — | — | — | 232.29 | 1 | E | Plant growth regulator |
| Metam-sodium | 6734-80-1 | 1 | — | — | — | — | 58.07 | 1 | E | Herbicide |
| Metamitron | 41394-05-2 | 2 | 1027.72 | — | — | — | 176.82 | 1 | E | Herbicide |
| Metazachlor | 67129-08-2 | 2 | 2255.00 | — | — | — | 233.15 | 2 | E | Herbicide |
| Methabenzthiazuron | 18691-97-9 | 1 | — | — | — | — | 96.53 | 1 | E | Herbicide |
| Methylarsonic acid | 124-58-3 | 1 | — | — | — | — | 49.36 | 1 | E | Herbicide |
| Metobromuron | 3060-89-7 | 2 | 4036.00 | — | — | — | 137.93 | 1 | E | Herbicide |
| Metolachlor | 51218-45-2 | 1 | — | — | — | — | 241.81 | 1 | E | Herbicide |
| Metosulam | 139528-85-1 | 2 | 2125.00 | — | — | — | 220.74 | 2 | E | Herbicide |
| Metribuzin | 21087-64-9 | 2 | 459.86 | — | — | — | 42.01 | 2 | E | Herbicide |
| Metsulfuron | 79510-48-8 | 1 | — | — | — | — | 241.81 | 1 | E | Herbicide |
| Metsulfuron-methyl | 74223-64-6 | 2 | 2510.00 | — | — | — | 261.19 | 2 | E | Herbicide |
| Molinate | 2212-67-1 | 1 | — | — | — | — | 214.16 | 1 | E | Herbicide |
| Monolinuron | 1746-81-2 | 3 | 1690.00 | — | — | — | 181.27 | 2 | E | Herbicide |
| Monosodium methyl-arsonate | 2163-80-6 | 1 | — | — | — | — | 96.86 | 1 | E | Herbicide |
| Napropamide | 15299-99-7 | 1 | — | — | — | — | 78.03 | 1 | E | Herbicide |
| Napthaleneacetic acid | 2122-70-5 | 2 | 2302.92 | — | — | — | 238.67 | 2 | E | Growth regulator |
| Nicosulfuron | 111991-09-4 | 2 | 2000.00 | — | — | — | 208.12 | 2 | E | Herbicide |
| Nonanoic acid | 112-05-0 | 1 | — | — | — | — | 261.32 | 1 | E | Herbicide, plant growth regulator |
| Norflurazon | 27314-13-2 | 2 | 1292.15 | — | — | — | 130.98 | 2 | E | Herbicide |

Table 3. (Continued).

| Chemical | CAS_RN | N | Median | Slope | Intercept | p | HD$_5$(50%) | #_EF | Status | Use |
|---|---|---|---|---|---|---|---|---|---|---|
| Orbencarb | 34622-58-7 | 2 | 2000.00 | — | — | — | 208.12 | 2 | E | Herbicide |
| Oryzalin | 19044-88-3 | 2 | 503.35 | — | — | — | 52.38 | 2 | E | Herbicide |
| Oxabentrinil | 74782-23-3 | 3 | 2500.00 | — | — | — | 439.20 | 2 | E | Herbicide safener |
| Oxadiazon | 19666-30-9 | 2 | 2274.16 | — | — | — | 192.89 | 2 | E | Herbicide |
| Oxasulfuron | 144651-06-9 | 1 | — | — | — | — | 216.76 | 1 | E | Herbicide |
| Oxyfluorfen | 42874-03-3 | 1 | — | — | — | — | 614.58 | 1 | E | Herbicide |
| Paclobutrazol | 76738-62-0 | 2 | 4953.25 | — | — | — | 193.05 | 1 | E | Plant growth regulator |
| Paraquat dichloride | 1910-42-5 | 4 | 243.32 | 0.12 | 4.65 | 0.37 | 88.50 | — | E | Herbicide |
| Pebulate | 1114-71-2 | 1 | — | — | — | — | 192.68 | 1 | E | Herbicide |
| Pentoxazone | 110956-75-7 | 1 | — | — | — | — | 261.32 | 1 | E | Herbicide |
| Phenmedipham | 13684-63-4 | 3 | 3107.23 | — | — | — | 299.93 | 1 | E | Herbicide |
| Phenyl-indole-3-thiobutyrate | 85977-73-7 | 1 | — | — | — | — | 216.76 | 1 | U | Herbicide |
| Phthalanilic acid | 4727-29-1 | 2 | 10700.00 | — | — | — | 1032.82 | 1 | E | Plant growth regulator |
| Picloram | 1918-02-1 | 3 | 2927.91 | — | — | — | 528.69 | 1 | E | Herbicide |
| Picloram-potassium | 2545-60-0 | 3 | 2121.32 | — | — | — | 482.63 | 1 | E | Herbicide |
| Picloram TIPA | 6753-47-5 | 1 | — | — | — | — | 216.76 | 1 | U | Herbicide |
| Pretilachlor | 51218-49-6 | 1 | — | — | — | — | 965.25 | 1 | E | Herbicide |
| Primisulfuron-methyl | 86209-51-0 | 2 | 2150.00 | — | — | — | 223.73 | 2 | E | Herbicide |
| Prodiamine | 29091-21-2 | 1 | — | — | — | — | 261.32 | 1 | E | Herbicide |
| Prohexadione-calcium | 127277-53-6 | 2 | 2000.00 | — | — | — | 208.12 | 2 | E | Plant growth regulator |
| Prometon | 1610-18-0 | 1 | — | — | — | — | 262.95 | 1 | E | Herbicide |
| Propachlor | 1918-16-7 | 1 | — | — | — | — | 10.57 | 1 | E | Herbicide |

Table 3. (Continued).

| Chemical | CAS_RN | N | Median | Slope | Intercept | $p$ | HD$_5$(50%) | #_EF | Status | Use |
|---|---|---|---|---|---|---|---|---|---|---|
| Propanil | 709-98-8 | 1 | — | — | — | — | 23.34 | 1 | E | Herbicide |
| Propaquizafop | 111479-05-1 | 2 | 2099.00 | — | — | — | 218.18 | 2 | E | Herbicide |
| Propham | 122-42-9 | 1 | — | — | — | — | 193.05 | 1 | E | Herbicide |
| Propisochlor | 86763-47-5 | 2 | 1344.00 | — | — | — | 66.41 | 1 | E | Herbicide |
| Propyzamide | 23950-58-5 | 2 | 5733.87 | — | — | — | 733.08 | 1 | E | Herbicide |
| Prosulfocarb | 52888-80-9 | 3 | 1650.00 | — | — | — | 232.82 | 1 | E | Herbicide |
| Prosulfuron | 94125-34-5 | 2 | 1622.00 | — | — | — | 159.59 | 2 | E | Herbicide |
| Pyrazosulfuron-ethyl | 93697-74-6 | 1 | — | — | — | — | 261.32 | 1 | E | Herbicide |
| Pyridate | 55512-33-9 | 1 | — | — | — | — | 160.51 | 1 | E | Herbicide |
| Pyriminobac-methyl | 136191-56-5 | 1 | — | — | — | — | 232.29 | 1 | E | Herbicide |
| Pyrithiobac-sodium | 123343-16-8 | 1 | — | — | — | — | 185.71 | 1 | E | Herbicide |
| Quinclorac | 84087-01-4 | 2 | 1974.68 | — | — | — | 205.46 | 2 | E | Herbicide |
| Quinmerac | 90717-03-6 | 1 | — | — | — | — | 232.29 | 1 | E | Herbicide |
| Quizalofop | 76578-12-6 | 2 | 2000.00 | — | — | — | 208.12 | 2 | E | Herbicide |
| Quizalofop-ethyl | 76578-14-8 | 2 | 2000.00 | — | — | — | 208.12 | 2 | U | Herbicide |
| Quizalofop-P-ethyl | 100646-51-3 | 2 | 2000.00 | — | — | — | 208.12 | 2 | E | Herbicide |
| Quizalofop-P-tefuryl | 119738-06-6 | 2 | 2150.00 | — | — | — | 223.73 | 2 | E | Herbicide |
| Rimsulfuron | 122931-48-0 | 2 | 1623.16 | — | — | — | 160.75 | 2 | E | Herbicide |
| Sebuthylazine | 7286-69-3 | 2 | 2732.05 | — | — | — | 334.37 | 1 | S | Herbicide |
| Sethoxydim | 74051-80-2 | 2 | 3755.00 | — | — | — | 482.63 | 1 | E | Herbicide |
| Siduron | 1982-49-6 | 1 | — | — | — | — | 2584.20 | 1 | E | Herbicide |
| Simazine | 122-34-9 | 2 | 7500.00 | — | — | — | 965.25 | 1 | E | Herbicide |
| Sodium arsenite | 7784-46-5 | 5 | 48.00 | 1.05 | -1.89 | 0.14 | 0.55 | — | U | Herbicide, insecticide |
| Sodium dimethylarsinate | 124-65-2 | 1 | — | — | — | — | 261.32 | 1 | E | Herbicide |
| Sulcotrione | 99105-77-8 | 2 | 1800.00 | — | — | — | 181.36 | 2 | E | Herbicide |
| Sulfentrazone | 122836-35-5 | 1 | — | — | — | — | 261.32 | 1 | E | Herbicide |

Table 3. (Continued).

| Chemical | CAS_RN | N | Median | Slope | Intercept | p | HD$_5$(50%) | #_EF | Status | Use |
|---|---|---|---|---|---|---|---|---|---|---|
| Sulfometuron-methyl | 74222-97-2 | 1 | — | — | — | — | 481.70 | 1 | E | Herbicide |
| TCA-sodium | 650-51-1 | 1 | — | — | — | — | 216.71 | 1 | E | Herbicide |
| Tebuthiuron | 34014-18-1 | 2 | 750.00 | — | — | — | 73.58 | 2 | E | Herbicide |
| Terbacil | 5902-51-2 | 1 | — | — | — | — | 262.37 | 1 | E | Herbicide |
| Terbuthylazine | 5915-41-3 | 2 | 1292.15 | — | — | — | 130.98 | 2 | E | Herbicide |
| Tetradec-11-en-1-yl acetate | 20711-10-8 | 1 | — | — | — | — | 249.71 | 1 | E | Herbicide |
| Thenylchlor | 96491-05-3 | 1 | — | — | — | — | 232.29 | 1 | E | Herbicide |
| Thiazafluron | 25366-23-8 | 1 | — | — | — | — | 25.58 | 1 | S | Herbicide |
| Thiazopyr | 117718-60-2 | 1 | — | — | — | — | 222.18 | 1 | E | Herbicide |
| Thidiazuron | 51707-55-2 | 1 | — | — | — | — | 367.02 | 1 | E | Plant growth regulator |
| Thifensulfuron | 79277-67-1 | 1 | — | — | — | — | 241.81 | 1 | E | Herbicide |
| Thifensulfuron-methyl | 79277-27-3 | 2 | 2510.00 | — | — | — | 261.19 | 2 | E | Herbicide |
| Thiobencarb | 28249-77-6 | 1 | — | — | — | — | 225.09 | 1 | E | Herbicide |
| Tiocarbazil | 36756-79-3 | 2 | 10000.00 | — | — | — | 1062.70 | 1 | E | Herbicide |
| Tralkoxydim | 87820-88-0 | 2 | 3725.00 | — | — | — | 290.94 | 1 | E | Herbicide |
| Tri-allate | 2303-17-5 | 1 | — | — | — | — | 261.44 | 1 | E | Herbicide |
| Triapenthenol | 76608-88-3 | 1 | — | — | — | — | 482.63 | 1 | S | Plant growth regulator |
| Triasulfuron | 82097-50-5 | 2 | 2150.00 | — | — | — | 223.73 | 2 | E | Herbicide |
| Tribenuron | 106040-48-6 | 1 | — | — | — | — | 261.32 | 1 | E | Herbicide |
| Tribenuron-methyl | 101200-48-0 | 1 | — | — | — | — | 261.32 | 1 | E | Herbicide |
| Tribufos | 78-48-8 | 3 | 273.00 | — | — | — | 51.13 | 2 | E | Plant growth regulator |
| Triclopyr BEE | 64700-56-7 | 1 | — | — | — | — | 91.76 | 1 | U | Herbicide |

Table 3. (Continued).

| Chemical | CAS_RN | N | Median | Slope | Intercept | p | HD$_5$(50%) | #_EF | Status | Use |
|---|---|---|---|---|---|---|---|---|---|---|
| Trifluralin | 1582-09-8 | 4 | — | — | — | — | 245.55 | 1 | E | Herbicide |
| Triflusulfuron | 135990-29-3 | 2 | 2250.00 | — | — | — | 234.13 | 2 | E | Herbicide |
| Triflusulfuron-methyl | 126535-15-7 | 2 | 2250.00 | — | — | — | 234.13 | 2 | E | Herbicide |
| Trinexapac-ethyl | 95266-40-3 | 1 | — | — | — | — | 192.68 | 1 | E | Plant growth regulator |
| Uniconazole | 83657-17-4 | 2 | 1888.00 | — | — | — | 191.37 | 2 | E | Plant growth regulator |
| Vermolate | 1929-77-7 | 2 | 3445.00 | — | — | — | 336.22 | 2 | E | Herbicide |
| CHEMICALS PRIMARILY ACTIVE AGAINST VERTEBRATES OR TESTED AS VERTEBRATE CONTROL AGENTS | | | | | | | | | | |
| [(3-Amino-2,4,6-trichloro phenyl) methylene]hydrazide | | 1 | — | — | — | — | 34.01 | 1 | U | Rodenticide |
| Benzenesulfonic acid (Bay 98663) 1,3-di-(fluorosulfonyl)cyclopentane (PHILLIPS 2133) | 35944-73-1 | 7 | 3.16 | 0.28 | -0.37 | 0.23 | 0.61 | — | U | Rodenticide |
| 1,3-Propanediol,2,2-bis(chloromethyl)-sulfate (PHILLIPS 2605) | 12712-28-6 | 6 | 8.75 | 0.38 | 0.11 | 0.29 | 0.90 | — | U | Tested as vertebrate agent |
| 3-Chloro-P-toluidine | | 10 | 22.45 | -0.02 | 2.88 | 0.97 | 0.30 | — | U | Avicide |
| 4-Aminopyridine (Avitrol) | 504-24-5 | 33 | 5.34 | -0.01 | 1.68 | 0.92 | 1.79 | — | U | Avicide |

Table 3. (Continued).

| Chemical | CAS_RN | N | Median | Slope | Intercept | $p$ | HD$_5$(50%) | #_EF | Status | Use |
|---|---|---|---|---|---|---|---|---|---|---|
| 5-(p-Chlorophenyl)-3,7,10-trimethyl sila-trane (D.M. 7537) | | 4 | 8.52 | 1.14 | -5.20 | 0.05 | 0.04 | — | U | Rodenticide |
| 6-Aminonicotinamide | | 1 | — | — | — | — | 0.77 | 1 | U | Rodenticide |
| Alpha-chloralose | 15879-93-3 | 18 | 69.16 | 0.26 | 3.12 | 0.07 | 14.05 | — | E | Rodenticide |
| Anthraquinone | 84-65-1 | 3 | 2000.00 | — | — | — | 193.05 | 1 | E | Bird Repellent |
| BAY COE 3664 | 39457-24-4 | 9 | 5.62 | 0.28 | 0.34 | 0.15 | 1.11 | — | U | Tested as vertebrate agent |
| BAY COE 3675 | 39457-25-5 | 9 | 2.37 | 0.69 | -2.03 | 0.00 | 0.38 | — | U | Tested as vertebrate agent |
| Brodifacoum | 56073-10-0 | 8 | 9.1 | -0.12 | 2.49 | 0.77 | 0.81 | — | E | Rodenticide |
| Bromadiolone | 28772-56-7 | 2 | 676.58 | — | — | — | 53.26 | 2 | E | Rodenticide |
| Bromethalin | 63333-35-7 | 1 | — | — | — | — | 0.83 | 1 | E | Rodenticide |
| Chlorophacinone | 3691-35-8 | 6 | 25.45 | -1.53 | 13.21 | 0.01 | 3.32 | — | E | Rodenticide |
| Cholecalciferol (vitamin D$_3$) | 67-97-0 | 1 | — | — | — | — | 192.68 | 1 | E | Rodenticide |
| Coumatetralyl | 5836-29-3 | 3 | 37.50 | — | — | — | 193.05 | 1 | E | Rodenticide |
| Difenacoum | 56073-07-5 | 1 | — | — | — | — | 9.32 | 1 | E | Rodenticide |
| Difethialone | 104653-34-1 | 3 | 0.87 | — | — | — | 0.31 | 2 | E | Rodenticide |
| Flocoumafen | 90035-08-8 | 4 | 51.25 | -0.64 | 6.45 | 0.50 | 0.07 | — | E | Rodenticide |
| Fluoroacetamide | 640-19-7 | 2 | 9.46 | — | — | — | 1.42 | 1 | E | Rodenticide |
| Methyl anthralinate | 134-20-3 | 2 | 1216.17 | — | — | — | 82.26 | 2 | U | Repellent |
| Metomidate | 5377-20-8 | 11 | 74.97 | 0.11 | 3.91 | 0.52 | 24.05 | — | U | Tested as vertebrate agent |
| Metomidate HCl | 35944-74-2 | 8 | 56.20 | 0.15 | 3.24 | 0.03 | 26.85 | — | U | Tested as vertebrate agent |

Table 3. (Continued).

| Chemical | CAS_RN | N | Median | Slope | Intercept | p | HD$_5$(50%) | #_EF | Status | Use |
|---|---|---|---|---|---|---|---|---|---|---|
| Pentobarbital-sodium | 57-33-0 | 8 | 107.66 | 0.00 | 4.71 | 0.97 | 48.93 | — | U | Soporific |
| Phencyclidine HCl | 956-90-1 | 13 | 75.00 | 0.14 | 3.63 | 0.57 | 9.32 | — | U | Tested as vertebrate agent |
| Phosacetim | 4104-14-7 | 12 | 16.61 | 0.19 | 2.00 | 0.68 | 0.41 | — | S | Rodenticide |
| Pindone | 83-26-1 | 1 | — | — | — | — | 27.99 | 1 | E | Rodenticide |
| Polyethoxylated aliphatic alcohols | 68131-40-8 | 3 | 2006.00 | — | — | — | 227.85 | 1 | U | Repellent |
| Scilliroside | 507-60-8 | 1 | — | — | — | — | 1.35 | 1 | S | Rodenticide |
| Sodium fluoroacetate (compound 1080) | 62-74-8 | 55 | 5.46 | 0.18 | 0.82 | 0.01 | 0.85 | — | E | Rodenticide |
| Sodium wafarin | 129-0601 | 2 | 1310.50 | — | — | — | 115.97 | 2 | U | Rodenticide |
| Starlicide | 7745-89-3 | 31 | 5.62 | -0.08 | 2.82 | 0.73 | 0.43 | — | U | Avicide |
| Strychnine | 57-24-9 | 17 | 6.00 | 0.15 | 1.47 | 0.36 | 1.04 | — | E | Rodenticide |
| Terrtiary butylsulfenyl-di-methyl dithiocarbamate | | 1 | — | — | — | — | 108.01 | 1 | U | Rodenticide |
| TFM (4-Nitro-3-[trifluoromethyl]phenol) | 88-30-2 | 1 | — | — | — | — | 44.12 | 1 | U | Lampricide |
| Thallium sulfate | 7446-18-6 | 4 | 40.48 | 0.10 | 3.10 | 0.70 | 8.93 | — | U | Rodenticide |
| Warfarin | 81-81-2 | 3 | 970.57 | — | — | — | 120.21 | 2 | E | Rodenticide |
| Zinc phosphide | 1314-84-7 | 7 | 44.85 | -0.34 | 5.40 | 0.15 | 5.45 | — | E | Rodenticide |

Table 3. (Continued).

**CHEMICALS WITH UNKNOWN SPECTRUM OF ACTIVITY**

| Chemical | CAS_RN | N | Median | Slope | Intercept | $p$ | HD$_5$(50%) | #_EF | Status | Use |
|---|---|---|---|---|---|---|---|---|---|---|
| Brofenprox | | 2 | 1942.00 | — | — | — | 201.99 | 2 | E | Unknown |
| Chloretazate | | 2 | 2150.00 | — | — | — | 207.53 | 1 | U | Unknown |
| Chloromethylmercury | 115-09-3 | 1 | — | — | — | — | 1.74 | 1 | U | Unknown |
| Cloxynil-sodium | | 1 | — | — | — | — | 3.72 | 1 | U | Unknown |
| Dibromoitrilopropi-onamide | | 2 | 184.11 | — | — | — | 18.96 | 2 | U | Unknown |
| Dichlorprop (racemic acid) | 7547-66-2 | 1 | — | — | — | — | 48.65 | 1 | E | Unknown |
| Flumequine | 42835-25-6 | 2 | 2500.00 | — | — | — | 96.53 | 1 | U | Unknown |
| Fluprimidol | | 1 | — | — | — | — | 232.29 | 1 | U | Unknown |
| Hexachlorbenzol | | 1 | — | — | — | — | 482.63 | 1 | U | Unknown |
| Propenamide | 79-06-1 | 1 | — | — | — | — | 19.26 | 1 | U | Unknown |

Table 4. List of pesticides currently in use with $HD_5(50\%)$ below 1 mg/kg.

| Compound | CAS_RN | n | Median | Slope | Intercept | p | $HD_5(50\%)$ | #_EF | Use Pattern |
|---|---|---|---|---|---|---|---|---|---|
| **CHOLINESTERASE-INHIBITING INSECTICIDES** | | | | | | | | | |
| Thiofanox | 39196-18-4 | 3 | 1.20 | na | na | na | 0.12 | 1 | |
| Terbufos | 13071-79-9 | 5 | 9.48 | 1.0277 | −1.5328 | 0.32 | 0.16 | — | |
| Propaphos | 7292-16-2 | 1 | — | na | na | na | 0.18 | 1 | |
| Carbofuran | 1563-66-2 | 18 | 1.65 | 0.0423 | 0.257 | 0.82 | 0.21 | — | |
| Phorate | 298-02-2 | 8 | 7.06 | 0.1817 | 0.4833 | 0.65 | 0.34 | — | |
| Parathion | 56-38-2 | 19 | 5.62 | 0.0797 | 1.2228 | 0.76 | 0.40 | — | |
| Quinalphos | 13593-03-8 | 2 | 20.65 | na | na | na | 0.42 | 1 | |
| Dicrotophos | 141-66-2 | 15 | 2.83 | 0.1787 | 0.3645 | 0.32 | 0.42 | — | |
| Monocrotophos | 6923-22-4 | 23 | 2.51 | −0.0312 | 1.0218 | 0.79 | 0.42 | — | |
| Aldicarb | 116-06-3 | 10 | 2.82 | 0.2955 | −0.6559 | 0.12 | 0.43 | — | |
| Fenamiphos | 22224-92-6 | 5 | 1.10 | −0.0863 | 0.5444 | 0.66 | 0.43 | — | |
| Isofenphos | 25311-71-1 | 6 | 10.96 | 0.0994 | 2.7255 | 0.86 | 0.44 | — | |
| Famphur | 52-85-7 | 3 | 2.70 | na | na | na | 0.45 | 1 | |
| Isazofos | 42509-80-8 | 3 | 11.10 | na | na | na | 0.51 | 2 | |
| EPN | 2104-64-5 | 14 | 6.43 | 0.3624 | 0.604 | 0.33 | 0.53 | — | |
| Diazinon | 333-41-5 | 14 | 5.25 | −0.2608 | 3.5883 | 0.29 | 0.59 | — | |

P. Mineau et al.

Table 4. (Continued).

| Compound | CAS_RN | n | Median | Slope | Intercept | p | HD$_5$(50%) | #_EF | Use Pattern |
|---|---|---|---|---|---|---|---|---|---|
| Coumaphos | 56-72-4 | 12 | 6.78 | 0.2179 | 0.7579 | 0.36 | 0.69 | — | |
| Mevinphos | 7786-34-7 | 13 | 3.80 | 0.0254 | 0.9409 | 0.88 | 0.70 | — | |
| Bendiocarb | 22781-23-3 | 4 | 16.24 | -0.9475 | 8.3724 | 0.37 | 0.72 | 2 | |
| Oxamyl | 23135-22-0 | 3 | 4.18 | na | na | na | 0.78 | — | |
| Disulfoton | 298-04-4 | 7 | 11.90 | 0.2019 | 1.2104 | 0.60 | 0.81 | 1 | |
| Cyanophos | 2636-26-2 | 1 | — | na | na | na | 0.83 | — | |
| Fenthion | 55-38-9 | 23 | 5.62 | 0.2581 | 0.4784 | 0.07 | 0.87 | 1 | |
| Triazamate | 112143-82-5 | 1 | — | — | — | — | 0.93 | — | |
| **NONCHOLINESTERASE INHIBITORS** | | | | | | | | | |
| Phenylmercuric acetate (PMA) | 62-38-4 | 4 | 145.86 | 2.19 | -9.43 | 0.16 | 0.01 | — | Fungicide |
| Flocoumafen | 90035-08-8 | 4 | 51.25 | -0.64 | 6.45 | 0.50 | 0.07 | — | Rodenticide |
| Chlordane | 57-74-9 | 4 | 62.28 | 1.00 | -1.63 | 0.44 | 0.09 | — | Insecticide |
| Difethialone | 104653-34-1 | 3 | 0.87 | — | — | — | 0.31 | 2 | Rodenticide |
| Bensultap | 17606-31-4 | 4 | 192.00 | 1.35 | -2.11 | 0.23 | 0.41 | — | Insecticide |
| Chlorfenapyr | 122453-73-0 | 2 | 8.30 | — | — | — | 0.56 | 1 | Insecticide, acaricide |
| Fenfuram | 24691-80-3 | 1 | — | — | — | — | 0.69 | 1 | Fungicide |
| Brodifacoum | 56073-10-0 | 8 | 9.1 | -0.12 | 2.49 | 0.77 | 0.81 | — | Rodenticide |
| Bromethalin | 63333-35-7 | 1 | — | — | — | — | 0.83 | 1 | Rodenticide |
| Sodium fluoroacetate | 62-74-8 | 55 | 5.46 | 0.18 | 0.82 | 0.01 | 0.85 | — | Rodenticide |

## III. Results and Discussion

Cholinesterase-inhibiting pesticides are grouped together in Table 2; all others are in Table 3. Pesticides are ordered alphabetically by their common chemical names except when a trade name only is available. A few trade names and synonyms are given to facilitate identification. All names follow the 11th Edition of the *Pesticide Manual* (Tomlin 1997) supplemented by the Nanogen Index (Packer 1975; Walker 1989). Tomlin (1997) was also used to determine whether any given product is still in commerce or is thought to no longer be marketed (superseded entries).

One advantage of this process is that it allows us to identify which pesticides currently in use are most toxic to birds and to start using some of the better known products for which field studies or incident records exist as possible "benchmarks" for equally toxic but more poorly known chemicals. From Tables 2 and 3, we arbitrarily selected those pesticides with an $HD_5(50\%)$ less than 1 mg/kg (Table 4). Of the 34 pesticides identified, 24 are cholinesterase-inhibiting insecticides. Of the remaining 10 products, 2 are insecticides including the very new pyrrole insecticide chlorfenapyr, 2 are fungicides, and the rest are rodenticides, including 3 of the second-generation coumarin anticoagulant products. Interestingly, the cholinesterase inhibitor thought to be the most toxic to birds is thiofanox (trade name Dacamox). According to the *Pesticide Manual* (Tomlin 1997), this granular and seed treatment insecticide is of only moderate toxicity to birds with $LD_{50}$ values of 109 mg/kg and 43 mg/kg in the mallard and the bobwhite, respectively. However, this is one of those few cases where this data source is in error. The values cited by Tomlin (1997) are in fact dietary $LC_{50}$ values; the true acute toxicity values are lower by 1.5 to 2 orders of magnitude.

In an early review of bird-kill incidents by Grue and colleagues (1983), these authors found that most incidents could be explained on the basis of pesticide toxicity and the extent to which the pesticides were used in U.S. agriculture. In a recent review of raptor incidents (Mineau et al. 1999), the higher proportion of kills resulting from labeled uses of pesticides in Canada and the U.S. relative to the U.K. was determined to be a result of the more permissive use of pesticides highly toxic to birds in North America. Certainly, those individuals familiar with pesticide bird-kill incidents throughout the world will recognize a number of compounds from Table 4 that keep coming back with depressing regularity. We believe that we have laid the groundwork for a more comprehensive review of those pesticides most hazardous to wild birds and for a fair comparison between older products and newer replacements. Recognizing that there are very few uses of pesticides that do not result in exposure to birds, we urge regulatory authorities to consider avian acute toxicity more closely before making regulatory decisions.

## Acknowledgments

We thank all the individuals, companies, and institutions who volunteered data for this effort. We also thank Charles Benbrook, and the Consumers Union as well as the W. Alton Jones Foundation for the support necessary to perform the time-consuming vetting of the database. A number of individuals demonstrated an infinite degree of patience with this task, particularly Rob Kriz and Lynn Schirml. We are grateful to Tom Aldenberg and Rick Bennett for comments on an earlier draft of this paper. Over the years, we benefitted from many a discussion with colleagues from around the world on this subject and, without being able to thank them all, would like to acknowledge at least the OECD, SETAC, and the USEPA for holding many of the forums where these discussions took place.

Appendix 1. Comparison of the equations for the expected value and variance from a sample with and without a covariate for bird weight. (A Fortran program to compute an $HD_5$ with body weight as a covariate is available by writing to the authors.)

|  | Without covariate | With covariate |
|---|---|---|
| Estimated expected value | $\bar{y} = \sum y_i/n$ | $\hat{y}_0 = \bar{y} + b(x_0 - \bar{x})$ |
| Variance of estimated expected value | $\sigma^2 \left[ \dfrac{1}{n} \right]$ | $\sigma^2 \left[ \dfrac{1}{n} + \dfrac{(x_0 - \bar{x})^2}{\sum (x_i - \bar{x})^2} \right]$ |
| Variance estimate ($s^2$) | $\sum (y_i - \bar{y})^2/(n - 1)$ | $\sum (y_i - \hat{y}_i)^2/(n - 2)$ |
| Degrees of freedom | $n - 1$ | $n - 2$ |

$$b = \frac{\sum y_i(x_i - \bar{x})}{\sum (x_i - \bar{x})^2}$$

$$\hat{y}_i = \bar{y} + b(x_i - \bar{x})$$

$x_0$ is the logarithm of the weight of the bird which you want to protect
$\hat{y}_0$ is the expected value of the logarithm of the $LD_{50}$ for the species of weight $x_0$
$\sigma$ is the standard deviation of the original $\log(LD_{50})$ data set
$\sigma^2$ is the unknown true population variance of the original $\log(LD_{50})$ data set
$s^2$ is the estimator of the variance

# References

Aldenberg T, Slob W (1993) Confidence limits for hazardous concentrations based on logistically distributed NOEC toxicity data. Ecotoxicol Environ Saf 25:48–63.

Baril A, Mineau P (1996) A distribution-based approach to improving avian risk assessment (abstract). 17th Annual SETAC, Washington, DC.

Baril A, Jobin B, Mineau P, Collins BT. (1994) A consideration of inter-species variability in the use of the median lethal dose ($LD_{50}$) in avian risk assessment. Tech Rep Series 216 Canadian Wildlife Service Headquarters, Hull, Québec.

Benbrook CM, Groth E, Halloran JM, Hansen MK, Marquardt S (1996) Pest Management at the Crossroads. Consumers Union, Yonkers, NY.

Dunning JB (1993) CRC Handbook of Avian Body Masses. CRC Press, Boca Raton.

Fischer DL, Hancock GA (1997) Interspecies extrapolation of acute toxicity in birds: body size scaling vs. phylogeny (abstract) 18th Annual SETAC. San Francisco, CA.

Grolleau G, Caritez JL (1986). Toxicité par ingestion forcée, de différents pesticides pour la perdrix grise, *Perdix perdix* L. et la perdrix rouge, *Alectoris rufa* L. Gibier Faune Sauvage 3:185–196.

Grue CE, Fleming WJ, Busby DG, Hill EF (1983) Assessing hazards of organophosphate pesticides to wildlife. Trans N Am Wildl Nat Res Conf (Washington, DC) 48:200–220.

Hart ADM, Thompson HM (1995) Significance of regurgitation in avian toxicity tests. Bull Environ Contam Toxicol 54:789–796.

Hudson RH, Tucker RK, Haegele MA (1984) Handbook of Toxicity of Pesticides to Wildlife. No. 153. U.S. Dept. of the Interior, Fish and Wildlife Service, Washington, DC.

Joermann G (1991) Comparative toxicity of pesticides to birds. Nachrbl Dtsch, Pflanzenschutzd (Stuttg) 43:275–279.

Kooijman SALM (1987) A safety factor for $LC_{50}$ values allowing for differences in sensitivity among species. Water Res 21:269–276.

Luttik R, Aldenberg T (1995) Extrapolation factors to be used in case of small samples of toxicity data (with special focus on $LD_{50}$ values for birds and mammals). RIVM Report. National Institute of Public Health and Environment. Bilthoven, The Netherlands.

Luttik R, Aldenberg T (1997) Extrapolation factors for small samples of pesticide toxicity data: special focus on $LD_{50}$ values for birds and mammals. Environ Toxicol Chem 16:1785–1788.

Mineau P (1991) Difficulties in the regulatory assessment of cholinesterase-inhibiting insecticides. In: Mineau P (ed) Cholinesterase Inhibiting Insecticides. Elsevier, Amsterdam, pp 277–299.

Mineau P, Jobin B, Baril A (1994) A critique of the avian 5-day dietary test ($LC_{50}$) as the basis of avian risk assessment. Tech Rep No. 215. Canadian Wildlife Service Headquarters, Hull, Québec.

Mineau P, Collins BT, Baril A (1996) On the use of scaling factors to improve interspecies extrapolation of acute toxicity in birds. Regul Toxicol Pharmacol 24:24–29.

Mineau P, Fletcher MR, Glaser LC, Thomas NJ, Brassard C, Wilson LK, Elliott JE, Lyon L, Henny CJ, Bollinger T, Porter SL (1999) Poisoning of raptors with organophosphorus and carbamate pesticides with emphasis on Canada, U.S., and U.K. J Raptor Res 33:1–37.

Newman MC, Ownby DR, Mezin LCA, Powell DC, Christensen TRL, Lerberg SB, Anderson BA (2000) Applying species-sensitivity distributions in ecological risk assessment: assumptions of distribution type and sufficient number of species. Environ Toxicol Chem 19:508–515.

Packer K (1975) Nanogen Index: A Dictionary of Pesticides and Chemical Pollutants. Nanogens International, Freedom, CA.

Sample BE, Arenal CA (1999) Allometric models for interspecies extrapolation of wildlife toxicity data. Bull Environ Contam Toxicol 62:653–663.

Schafer EW, Brunton RB (1979) Indicator bird species for toxicity determinations: is the technique usable in test method development? In: Beck JR (ed) Vertebrate Pest Control and Management Materials. American Society for Testing and Materials, Philadelphia, pp 157–168.

Schafer EW Jr, Bowles WA Jr, Hurlbut J (1983) The acute oral toxicity, repellancy and hazard potential of 998 chemicals to one or more species of wild and domestic birds. Arch Environ Contam Toxicol 12:355–382.

SETAC (1996) Report of the SETAC/OECD Workshop on Avian Toxicity Testing. OECD, Paris.

Smith GJ (1987) Pesticide Use and Toxicology in Relation to Wildlife: Organophosphorus and Carbamate Compounds. U.S. Fish and Wildlife Service, Washington, DC.

Stephan CE, Rogers JW (1985) Advantages of using regression analysis to calculate results of chronic toxicity tests. In: Bahner RC, Hansen DJ (eds) Aquatic Toxicology and Hazard Assessment: 8th Symposium. STP 891. American Society for Testing and Materials, Philadelphia, pp 328–338.

Tomlin CDS (1997) The Pesticide Manual, 11th Ed. British Crop Protection Council, Binfield, Berkshire, UK.

Tucker RK, Haegele MA (1971) Comparative acute oral toxicity of pesticides to six species of birds. Toxicol Appl Pharmacol 20:57–65.

Van Straalen NM, Denneman CAJ (1989) Ecotoxicological evaluation of soil quality criteria. Ecotoxicol Environ Saf 18:241–251.

Wagner C, Lokke H (1991) Estimation of ecotoxicological protection levels from NOEC toxicity data. Water Res 25:1237–1242.

Walker SB (1989) Consolidated Index for the Updated Nanogen Index. Nanogens International, Freedom, CA.

Manuscript received March 17, 2000; accepted July 10, 2000.

Rev Environ Contam Toxicol 170:75–140

# Molecular Chlorine: Health and Environmental Effects

## Karen M. Vetrano

## Contents

## I. Introduction

This review is intended to provide information regarding the health and environmental effects resulting from exposure to chlorine and was prepared by TRC Environmental Corporation on behalf of The Chlorine Institute. The Chlorine Institute is a Chemical Manufacturers Association (CMA) Responsible Care® Partnership Association. In this capacity, the Institute is committed to fostering the adoption by its members of the Codes of Management Practices; facilitating their implementation; and encouraging members to join the Responsible Care® initiative directly.

Communicated by George W. Ware.

K.M. Vetrano
TRC Environmental Corporation, 5 Waterside Crossing, Windsor, CT 06095, U.S.A.

The purpose of this review is to present the findings of published research related to the toxicity and environmental effects of chlorine gas. This report does not provide a discussion of the by-products of water chlorination (drinking water and wastewater). For a discussion of the water chlorination issue, please refer to The Chlorine Institute's 1990(a) "Toxicity Summary for Chlorine and Hypochlorites, and Chlorine in Drinking Water" and the 1991 *IARC Monographs on the Evaluation of Carcinogenic Risks to Humans*, Volume 52.

The scope of this study is as follows:

1. To review the toxic potential of chlorine by exposure following inhalation, dermal and eye contact, or ingestion by humans and animals.
2. To review the effects of chlorine on the environment.
3. To provide an overview of governmental and nongovernmental regulatory standards and guidelines.

All abbreviations used in this review are presented in the Appendix.


## II. Overview of Chlorine

In 1979, world production of chlorine exceeded 35 million t, and by 1995, production increased to more than 43 million t of chlorine (Westervelt 1996). Chlorine Institute data show that 13,807,739 t of chlorine was produced in the United States in 1999. Chlorine production in Canada in 1997 was 1,117,000 t and in Mexico in 1997 (latest year available) 457,000 t (Chlorine Institute 2000).


### A. Physicochemical Properties

Chlorine is the most abundant and reactive of all halogens. Chlorine is an element found in nature only in the combined state, primarily with sodium as a common salt (NaCl) as well as with potassium and magnesium, such as carnallite ($KMgCl_3 \cdot 6H_2O$), and sylvite (KCl) (Laubusch 1962a). At standard conditions, chlorine exists as a yellowish-green gas. No information regarding the exact concentration at which chlorine gas becomes visible was found in the literature. At low temperatures or under high pressure, chlorine liquefies, and it is usually transported and stored as such.

One liter of liquid chlorine produces approximately 434 L of chlorine gas at 25 °C (Frank 1986). Because chlorine is approximately 2.5 times heavier than air, a chlorine cloud will tend to remain at the ground surface and dissipate slowly. Tables 1 and 2 summarize the physical and chemical properties of chlorine.

A more complete review of the physical and chemical properties of chlorine can be found in the following reviews: *Chlorine* (J.S. Sconce, ed. A.C.S. Monograph Series, 154. Reinhold, New York, 1962); and Downs AJ, Adams CJ, Chlorine, Bromine, Iodine and Astatine (in: Bailar JC, et al., eds.) *Comprehensive Inorganic Chemistry*, Vol. 2. Pergamon Press, Oxford, 1973).

Table 1. Physical and chemical properties of chlorine.[a]

| Chemical name, CAS number | Physical description | Chemical and physical properties | Incompatibilities and reactivities |
| --- | --- | --- | --- |
| Chlorine, $Cl_2$ 7782-50-5 | Yellowish-green gas with pungent, irritating odor<br><br>Shipped as a lique-fied, compressed gas | MW: 70.9<br>BP: −29.29 °F<br>  (−34.05 °C)<br>  239.05 K[c]<br>MP: −101.5 °C[b]<br>  172.12 K[c]<br>SG: 1.56 (at −33.6 °C)[b]<br>Sol.: 0.7%<br>Fl.P: NA<br>Density (vs. air): 2.47<br>VP: 6.8 atm<br>Crit. temp.: 144 °C[d]<br>Crit. press.: 76.1 atm[d]<br>FRZ: −150 °F<br>UEL: Not explosive<br>LEL: Not explosive<br>Note: Nonflammable gas, but strong oxidizer | Reacts explosively or forms explosive com-pounds with many com-mon substances such as acetylene, ether, turpen-tine, ammonia, fuel gas, hydrogen, and finely di-vided metals |

MW, molecular weight; BP, boiling point; MP, melting point; SG, specific gravity; Sol, solubility in water at 68 °F, % by weight (i.e., g/100 ml); Fl.P, flash point; NA, not available; VP, vapor pressure; FRZ, freezing point, °F; UEL, upper explosive limit in air; LEL, lower explosive limit in air.
[a]NIOSH (1994).
[b]Lide (1998).
[c]Giauque and Powell (1939).
[d]Budaveri (1996).

## B. Chlorine Uses

Chlorine is one of the most frequently used industrial chemicals. Industrial uses of chlorine fall into two general categories: (i) the production of various organic and inorganic chlorinated compounds (87% of total) and (ii) drinking water and wastewater disinfection (5%). Bleaching processes in the pulp and paper indus-try accounted for 8% in 1995; however, this use is gradually being phased out (Chemical Marketing Reporter 1995).

Most chlorine produced in the U.S. is used in the production of chlorinated organic and inorganic compounds. It is used in the manufacture of a variety of chemicals that are subsequently used in the manufacture of an extensive array of consumer products such as plastics and pharmaceuticals. Table 3 (p. 79) provides a listing of the chemical intermediates produced from chlorine and the resultant consumer products (Munro 1994).

The use of chlorine gas in water treatment includes disinfection of potable water, wastewater effluent treatment, and as a biocide for equipment maintenance in power generation plants, desalination plants, petrochemical plants, and the paint and metal industries (NAS 1976). The strong oxidizing nature of chlorine, its ease of application, and low cost have led to its use as a primary disinfectant. This disinfectant ability of free chlorine (HOCl/OCl⁻) is related to the oxidizing nature of the free chlorine residual such that it can inactivate a broad range of waterborne bacterial, viral, and protozoal pathogens. It is used in the treatment of sewage effluent to control pathogenic organisms, reduce odor, and reduce biochemical oxygen demand (BOD) (Laubusch 1962b). Chlorine is also used as a biocide to reduce biofouling in cooling water towers of power plants, desalination plants, and petrochemical, paint, and metal industries. Chlorination removes biota that form on conduits, piping, and the heat-exchange surfaces of condensers (NAS 1976). Circulating cooling water is warm, well oxygenated, and provides an ideal environment for waterborne growth. Untreated cooling systems are subject to fungal rot of wooden parts of the tower, bacterial corrosion of iron and bacterial production of sulfide, and large growths of algae in the sunlit portions of the tower. Biocidal chemicals must be added to control growth (USEPA 1977). Chlorine is used as a treatment for microfouling to remove slimes and algae and bacteria and its extracellular excretions or for

Table 2. Vapor pressure of chlorine (0 °C = 273.10 °K)

| T, K | T, °C | P, observed international cm Hg |
|---|---|---|
| 172.12 | −100.98 | 1.044 |
| 175.44 | −97.66 | 1.407 |
| 180.38 | −92.72 | 2.158 |
| 185.38 | −87.72 | 3.250 |
| 190.513 | −82.587 | 4.774 |
| 195.513 | −77.587 | 6.831 |
| 200.413 | −72.687 | 9.507 |
| 205.242 | −67.858 | 12.949 |
| 210 | −63.1 | 17.274 |
| 215.179 | −57.921 | 23.251 |
| 219.909 | −53.191 | 30.124 |
| 225.104 | −47.996 | 39.396 |
| 229.958 | −43.142 | 50.042 |
| 234.975 | −38.125 | 63.273 |
| 240.05 | −33.05 | 79.385 |

Adapted from Giaque and Powell (1939).

Table 3. Chemicals and products derived from chlorine. (Reproduced from Munro, 1996).

## Chemical intermediates

| | |
|---|---|
| Acetyl chloride | Ethylene dichloride |
| Allyl chloride | Hexachlorobenzene |
| Aluminum chloride | Hydrochloric acid |
| Benzoyl chloride | Methyl chloride |
| Benzyl chloride | Methyl parathion |
| 1-Butanol | Methylene chloride |
| Calcium hypochloride | Monochloroacetic acid |
| Carbon tetrachloride | Pentachlorophenol |
| Chloral | Perchloroethylene |
| Chloroacetic acid | Phosgene |
| Chlorobenzene | Phosphorus oxychloride |
| Chloroform | Phosphorus trichloride |
| Chloroparaffins, (C10–C30)Cl | Potassium hydroxide |
| Chloroprene | Sulfuryl chloride |
| Dichlorobenzene | 1,1,1-Trichloroethane |
| 1,2-Dichloropropane | Trichloroethylene |
| Dichlorophenol | Vinyl chloride |
| Epichlorohydrin | Vinylidene chloride |
| Ethyl chloride | |

## Consumer products

| | |
|---|---|
| Adhesives | Leather processing |
| Aeorosols | Metal processing |
| Anesthetics | Oil processing |
| Antifreeze | Paints/paint removers |
| Bactericides | Pesticides |
| Bleaches | Pharmaceuticals |
| Ceramics | Photographic film |
| Cosmetics | Moth proofing |
| Deodorants | Pigments |
| Detergents | Plastics |
| Disinfectants | Printing inks |
| Drinking water | Pulp and paper making |
| Dry cleaning | Refrigerants |
| Dyes | Rocket fuels |
| Electronics | Sanitizing agents |
| Explosives | Sewage treatment |
| Fertilizers | Specialty glass making |
| Fire extinguishers | Surfactants |
| Fire retardants | Swimming pool treatment |
| Food additives/food processing | Synthetic rubber |
| Fumigants | Textiles |
| Fungicides | Tobacco |
| Gasoline | Wood preservatives |
| Herbicides | Zirconium |
| Hydraulic fluids | |

macrofouling to eliminate hydroids, barnacles, mussels, clams, and oysters in marine cooling systems.

The pulp and paper industry, which at one time accounted for 18% of total chlorine consumption, uses chlorine to bleach paper pulp and to purify dissolving pulps to obtain high concentrations of alpha-cellulose (Chemical Marketing Reporter 1989). However, this use is being gradually phased out, with bleaching processes in the pulp and paper industry accounting for 8% of the chlorine used in 1995 (Chemical Marketing Reporter 1995). The four basic techniques used to obtain these results are acid chlorination in dilute solution, alkaline hypochlorite bleaching, caustic extraction, and chlorine dioxide bleaching (NAS 1976).

One of the miscellaneous uses of chlorine is in food processing. It is used in large quantities to wash meat carcasses, seafood, vegetables, and fruits. It has been shown that the immersion of chicken carcasses in an aqueous solution of chlorine can increase shelf life by approximately 20% (Patterson 1968). Chlorine is also used as a bleaching agent in flour to improve its cake-baking qualities (Tsen and Kulp 1971; Kulp et al. 1972). Chlorine solutions are also used to disinfect equipment surfaces during food processing.

## C. Atmospheric Chemistry of Chlorine

Both federal and state governments have regulatory programs in place dealing with atmospheric releases of chlorine. Industrial handling of chlorine is regulated by the USEPA under the Clean Air Act Amendments of 1990 as a Hazardous Air Pollutant (HAP) and as a toxic gas under the Risk Management Program. The atmospheric concentration of chlorine is not regulated by the USEPA. OSHA, under the Process Safety Management program, also regulates industrial handling of chlorine. Atmospheric concentrations of chlorine are regulated by many states under individual "air toxics" programs. A more comprehensive review of the atmospheric chemistry of chlorine can be found in The Chlorine Institute's Pamphlet 84, Environmental Fate of Chlorine in the Atmosphere (Chlorine Institute 1990b).

*Sources of Atmospheric Chlorine.* The most significant, natural sources of atmospheric chlorine are from the photolysis of salt in seawater and volcanic emissions. Duce (1969) identified sea salt as a source of chlorine; however, no estimates of the quantities of chlorine from such processes were made. It has since been concluded that sea spray, which releases hydrochloric acid (HCl) into the marine boundary layer (Singh and Kasting 1988), may be photochemically oxidized in the atmosphere to form chlorine (McConnel et al. 1992). The total quantities of chloride available in geologic formations and seawater have been estimated to be 1,000 and 36,000 billion t, respectively (Hopp 1981). The estimated ranges of chloride released (as HCl) from global volcanic eruptions have been reported to range from 0.3–11 million t by Symonds et al. (1988). Less significant natural sources of chlorine in the atmosphere include the production of organochlorides by plant and animal species (notably algae), production of chloromethane, as a product of metabolism, by fungi and other plants, and through the burning of chlorine containing vegetation in forest fires and controlled burns. These sources are minor when compared to anthropogenic sources. In its Toxic Release Inventory, the USEPA reported that 33,128 t of chlorine were released from U.S. industries in 1995 (USEPA 1995).

Environmentally Important Atmospheric Reactions.    The Earth's atmosphere can be thought of as a series of vertical layers, each of which is characterized by different chemical and physical properties. The atmospheric layer nearest the Earth's surface, which supports life and weather phenomena, is known as the troposphere. The troposphere extends from the surface to altitudes of about 10 km. The layer immediately above the troposphere is known as the stratosphere. The stratosphere is characterized by conditions of pressure, temperature, and density that are quite different than those in the troposphere. Because of these differences the chemical and physical processes that occur in the stratosphere are also quite different from those in the troposphere.

Ozone ($O_3$) is a molecule composed of three atoms of oxygen. There are environmental effects related to the occurrence of ozone in each of these atmospheric layers. The discussion of these effects can become confusing because the conditions that are associated with harmful effects appear to be contradictory. The three most probable reaction pathways for atmospheric chlorine are photolysis, hydroxyl reactions, and ozone reactions.

*Photolysis.*    The principal chlorine reaction pathway is photolysis. This pathway only occurs during daytime hours because it depends on sunlight for initiation. Chlorine molecules ($Cl_2$) react with sunlight (in the wavelength range 240–450 nm) to produce atomic chlorine (Cl) (see Eq. 1):

$$Cl_2 + hv \rightarrow 2Cl \tag{1}$$

where $Cl_2$ is molecular chlorine, hv is sunlight (240–450 nm), and Cl is atomic chlorine. The speed of this reaction varies with the intensity and wavelength of light; however, the estimated daytime half-life of $Cl_2$ is about 2–4 hr. Atomic chlorine may subsequently react with other atmospheric species including hydrocarbons, nitrogen oxides, and hydroxyl radicals ($OH^-$).

*Hydroxyl Reactions.*    Molecular chlorine can also react with tropospheric hydroxyl radicals forming atomic chlorine and hypochlorous and hydrochloric acid as shown in Eqs. 2 and 3:

$$Cl_2 + OH \rightarrow Cl + HOCl \tag{2}$$

and

$$Cl_2 + OH \rightarrow OCl^- + HCl \tag{3}$$

where $Cl_2$ is molecular chlorine, OH is hydroxyl radical, Cl is atomic chlorine, $OCl^-$ is hypochlorite ion, HOCl is hypochlorous acid, and HCl is hydrochloric acid. These reactions are much slower than photolysis but probably provide the major sink for atmospheric molecular chlorine at night (Loewenstein and Anderson 1984).

*Ozone Reactions.* Ozone occurs both in the troposphere and the stratosphere, the layer of the earth's atmosphere from an altitude of about 20 km to 75 km. In the troposphere, ozone is a major component of "smog" and is a "criteria pollutant" regulated by the USEPA. Ozone is formed in the troposphere through a complex chemical mechanism that depends on the emissions of oxides of nitrogen ($NO_x$) and volatile organic compounds (VOC) that result from many different anthropogenic and natural sources. Ozone is not a natural component of the troposphere, nor is it emitted directly from anthropogenic activities. Ozone is a strong oxidizing agent and has been associated with human health effects involving impairment of respiratory functions in exposed populations. Therefore, human exposure to ozone in the near-surface troposphere represents an environmental problem of considerable importance. Although significant research has been conducted to understand tropospheric ozone formation processes and to develop effective control strategies, tropospheric ozone concentrations continue to exceed air quality standards in many urban areas.

Ozone is a naturally occurring component of the stratosphere. The conditions in the stratosphere result in an equilibrium state between the atomic, diatomic, and triatomic forms of oxygen. The net result is a concentration of ozone that is maintained in equilibrium at natural conditions. The ozone that occurs naturally in the stratosphere absorbs ultraviolet radiation received from the sun. As such, the stratospheric ozone layer acts as an ultraviolet filter, and only small amounts of the incident ultraviolet radiation pass through to the troposphere. Because of the differences in pressure and density in the troposphere, the equilibrium condition cannot be maintained below the stratosphere. The ultraviolet energy that penetrates the stratospheric ozone layer has been linked to the formation of skin cancers and other health-related effects. Therefore, the depletion of stratospheric ozone also represents a serious environmental problem.

Natural atmospheric mixing can result in the injection of tropospheric gases into the stratosphere. The mixing time required for tropospheric stratospheric exchange is long and therefore, only the most stable of compounds can be mixed into the stratosphere in any large quantities. Molecular chlorine has a relatively short lifetime in the troposphere and therefore is not considered to be active in the stratospheric ozone depletion process.

Other chlorine-containing compounds, most notably the chlorofluorocarbons (CFCs), are essentially inert in the troposphere and therefore are eventually mixed into the stratosphere. When exposed to the energetic ultraviolet radiation available in the stratosphere they can photodissociate, releasing atomic chlorine. The atomic chlorine can enter into reactions with the oxygen in the stratosphere. The presence of atomic chlorine can disrupt the stratospheric ozone equilibrium through a series of reaction steps. This process may result in a net depletion of the available ozone in the stratosphere and a subsequent increase in the amount of ultraviolet radiation that can pass through to the troposphere.

Atomic chlorine reacts with ozone through the following steps (McElroy and Salawitch 1989) (Eqs. 4a, 4b):

$$Cl + O_3 \rightarrow OCl^- + O_2 \tag{4a}$$

$$OCl^- + O \rightarrow Cl + O_2 \tag{4b}$$

where Cl is molecular chlorine, $O_3$ is ozone, and $OCl^-$ is hypochlorite ion. In this process, atomic chlorine destroys a molecule of ozone before it absorbs UV radiation, and atomic chlorine is regenerated to continue the reaction. This process shifts the stratospheric ozone equilibrium, resulting in less radiation-absorbing ozone and thus more UV radiation being able to penetrate to the earth's surface.

*Fate of Atmospheric Chlorine.*   Typical concentrations of chlorine atoms in the troposphere, the layer of the atmosphere from the earth's surface to an altitude of about 15–20 km, are of the order of 0.2–3 ppbv, with highest concentrations found in the tropical marine boundary layer and lowest concentrations found in continental air masses. Ultimately, chlorine likely forms hydrochloric (HCl) or hypochlorous (HOCl) acid in the atmosphere, either through reactions with hydroxyl radicals or other trace species, such as hydrocarbons. These acids are believed to be removed from the atmosphere primarily through precipitation washout (i.e., wet deposition as chlorine is scrubbed out by rain in the subcloud layer) or dry deposition as gaseous chlorine contacts and reacts with the earth's surface.

## D. Water Chemistry and Fate

Water chlorination, resulting from municipal and industrial wastewater treatment or cooling water disinfection, initially introduces chlorine into the water as chlorine gas, hypochlorite ion ($OCl^-$), or its salt. These forms of chlorine are termed free residual chlorine (FRC) (Mattice and Zittel 1976). Chlorine in aqueous systems volatilizes or quickly decays to residual forms such as chloramine. Aquatic chlorine chemistry is determined by other aquatic factors including pH, ammonium ion, which combines with chlorine to form chloramine, and certain other reducing agents. Inorganic reducing agents in estuarine water include sulfur, iron, and manganese. Other organic compounds in the water also contribute to chlorine decay rates (Stober and Hanson 1974). A measurement of total residual chlorine (TRC) includes FRC and combined chlorine, such as chloramines and other chlorinated organics in water (Mattice and Zittel 1976). The reactions of chlorine and hypochlorites in water produce a number of by-products, many of which have been implicated as genotoxic or tumorogenic. The functional behavior of chlorine gas and hypochlorites is similar when added to water. Therefore, the discussion in this report does not distinguish between the activity of these two agents in water.

*Reactivity of Chlorine in Water.* Chlorine gas added to water immediately forms the hypochlorite ion and hydrogen chloride from the dissociation of a water molecule in the following manner:

$$pH > 7.5$$
$$\longrightarrow$$
$$Cl_2 + H_2O \longrightarrow H^+ + Cl^- + HOCl \longleftarrow H^+ + OCl^- \tag{5}$$
$$pH < 7.5$$

Hypochlorites added to water yield the hypochlorite ion directly.

$$Ca(OCl)_2 + H_2O \longrightarrow Ca^{2+} + 2\,OCl^- + H_2O \tag{6}$$

As shown in Eq. 5, formation of the hypochlorous acid is favored at pH 7.5 or lower, and the hypochlorite ion is favored at higher pH values. At pH 3.5 or lower (not shown), the formation of elemental chlorine is favored. These forms of chlorine are referred to as free available chlorine. Subsequent reaction of free chlorine is dependent on pH, temperature, and the availability of other chemical constituents in the water, as well as their relative concentrations and equilibria constants.

Chlorine gas is rarely added to wastewater effluents; instead, disinfection of the effluent is achieved by preparing a concentrated "solution" of the gas in water, which is then added to the effluent. When chlorine is added to water containing natural or added ammonia, the ammonia reacts with HOCl to form various chloramines, depending on the pH of the solution and the initial $Cl_2$:$NH_3$ ratio. The following reactions occur (USEPA 1974) (Eqs. 6a–6d):

Reaction 1:  $Cl_2 + H_2O \Rightarrow HOCl + HCl$ $\qquad\qquad\qquad\qquad$ (6a)

Reaction 2:  $NH_3 + HOCl \Rightarrow NH_2Cl$ (monochloramine) $+ H_2O$ $\quad$ (6b)

Reaction 3:  $NH_2Cl + HOCl \Rightarrow NHCl_2$ (dichloramine) $+ H_2O$ $\quad$ (6c)

Reaction 4:  $NHCl_2 + HOCl \Rightarrow NCl_3$ (nitrogen trichloride) $+ H_2O$ $\quad$ (6d)

Whether one of these compounds or a combination of them is formed depends on the pH of the water and on an excess of ammonia. In general, low pH levels and high $Cl_2$:$NH_3$ ratios favor dichloramine formation. Above pH approximately 8.5, monochloramine exists almost exclusively. At pH between approximately 8.5 and 5, monochloramine and dichloramine exist simultaneously; at pH between approximately 5.5 and 4.5, dichloramine exists almost exclusively; and below pH about 4.4, nitrogen trichloride is produced (Laubusch 1962c). Chlorine in combination with ammonia nitrogen and organic nitrogen compounds is termed combined available chlorine. Chloramines are important in providing a long-lived "combined chlorine residual" in water distribution systems (Hammer 1975).

Reaction products in freshwater and saltwater are different. Table 4 lists the principal chlorine reactions of concern in aqueous solutions (Jolley and Carpenter 1984). In saline waters, the availability of bromine causes the rapid oxidation of HOCl to yield HOBr or OBr. Ammonia, a typical constituent of estuarine waters, reacts to form mono- and dibromoamines as well as monochloramine (Capuzzo et al. 1977).

As indicated in Table 4, chlorine also oxidizes organic compounds in water. Plant decomposition products leach into water and form the major source of organic compounds in natural waters (i.e., humic compounds). The substitution of chlorine into amino and aromatic compounds causes production of trihalomethanes (THM) and organic halogens (TOX), such as haloacetonitriles and chloroacetic acid. In industrial effluents, phenols provide a source for aromatic compounds leading to production of THM and TOX (Johnson and Jensen 1986). The nonvolatile, organic-bound chlorine produced by these reactions exceeds the volatile components such as chloroform by a factor of 3 to 5 (Jolley 1984). Generally, identifying the specific by-products of chlorination in natural waters has been unsuccessful (Johnson and Jensen 1986).

The sum of the chloramines and organo-$N$-chloro compounds are identified as combined available or combined residual chlorine. Total residual chlorine (TRC) is the sum of both free and combined forms and is the most frequently cited measure of chlorine concentrations in water. In seawater, the participation of bromine, and to a lesser extent iodine, in the previously described reactions has necessitated measuring total residual oxidant (TRO) in seawater to include the contribution of chlorine, bromine, and iodine.

*Fate of Chlorine in Water.* Chlorine's ultimate aquatic fate is chloride (Jolley and Carpenter 1984). Chloride from chlorination processes is relatively insignificant compared to background levels (Stevens et al. 1985). The oxidation of organic materials (e.g., aldehyde oxidation to carboxylic acid with the conse-

Table 4. Principal reactions of chlorine in water.

| Reaction | Example |
|---|---|
| Water | $Cl_2 + H_2O \leftrightarrow HOCl + HCl$ |
| Ammonia | |
|     Substitution | $NH_3 + HOCl \rightarrow NH_2Cl + H_2O$ |
|     Oxidation | $2NHCl_2 + H_2O \rightarrow N_2 + HOCl + 3H^+ + 3Cl^-$ |
| Inorganic oxidation | $Mn^{2+} + HOCl + 2H_2O \rightarrow MnO(OH)_2 + 3H^+ + Cl^-$ |
| Disproportionation | $3OCl^- \rightarrow 2Cl^- + ClO^-$ |
| Decomposition | $2HOCl \rightarrow 2H^+ + 2Cl^- + O_2$ |
| Organics | |
|     Oxidation | $RCHO + HOCl \rightarrow RCOOH + H^+ + Cl^-$ |
|     Addition | $RC \leftrightarrow CR' + HOCl \rightarrow RC(OH)C(Cl)R'$ |
|     Substitution | |
|         N–Cl bond | $RNH_2 + HOCl \rightarrow RNHCl + H_2O$ |
|         C–Cl bond | $RCOCH_3 + 3HOCl \rightarrow RCOOH + HCCl_3 + 2H_2O$ |

quent reduction of HOCl to chloride) is the most important process in the decay of free chlorine in the environment (Dotson et al. 1986). Chlorination of waters finally leads to chloride, oxidized organics ($CO_2$), chloro-organics, $O_2$, and $N_2$.

## III. Health Effects of Chlorine

The information gathered on the health effects associated with chlorine includes data collected from animal studies and human volunteers, case studies of accidental exposures, and epidemiological studies of exposed populations. The data collected from animal studies include short- and long-term exposures at various concentrations. Data derived from controlled voluntary human exposures are associated with low concentrations and limited time periods. Case studies of accidental exposures provide information on the clinical outcome of chlorine exposure; however, the exposure concentration is often not known. Epidemiological case studies of exposed populations, primarily occupational exposures, provide a means to evaluate long-term exposures to humans at low concentrations. However, the relevance of these studies is often hampered by confounding factors (e.g., differences in age, smoking habits, etc., of the exposed population), which can complicate the interpretation of the study findings (e.g., pulmonary function deficits from chronic chlorine inhalation or as a result of chronic smoking).

### A. Human Exposure

Chlorine, regardless of its source and use, is found in only two states: gas and liquid. The most common state of chlorine in accidental exposures is the gaseous state. Therefore, the most important route of exposure is inhalation, followed by eye and skin exposures. Chlorine gas was used as a chemical warfare agent near Ypres, where the Germans first used chlorine gas against French Colonial troops in 1915 (Sandall 1922; Penington 1954; Anonymous 1984). Modern catastrophic exposures to chlorine gas primarily result from storage or transportation accidents involving the pressurized liquid form. Other incidents have occurred in industrial accidents, school chemistry experiments, and accidental releases from swimming pool operations and mixing of household cleaning agents, (e.g., adding acidic cleaning agents to bleach releases chlorine gas). In addition, employees of facilities that use chlorine in their day-to-day operations may be exposed to low levels of chlorine.

*Pharmacokinetics.*    There is limited published literature regarding the pharmacokinetics of inhaled chlorine gas. Nodelman and Ultman (1999a,b) recently evaluated the distribution of chlorine absorption in human airways after nasal and oral breathing. Nodelman et al. (1998) devised a fast-responding thermionic chlorine analyzer to noninvasively determine the distribution of chlorine in the intact human respiratory tract. Using this analyzer, Nodelman and Ultman measured the longitudinal distribution of chlorine gas by the bolus inhalation method in 10 healthy nonsmokers during nasal and oral breathing during quiet breathing (1999a) and at respiratory flows of 150, 250, and 1000 mL/sec (1999b). In the first set of experiments, healthy nonsmoking subjects were administered 0.5 and 3 ppm chlorine gas boluses during quiet breathing (Nodelman and Ultman 1999a). During this set of experiments it was determined that nearly all the chlorine gas inhaled during quiet breathing was absorbed by the upper airways, regardless of administration via the nose or through the mouth, and that the total absorption rates for the nose and mouth are similar. Nodelman and Ultman also determined that the dissolution, diffusion, and chemical reactions governing chlorine gas uptake from respired gas to the nasal mucosa are all linear processes, because the absorption constant in the nasal–oropharynx region stayed the same when $C_{max}$ was changed from 0.5 to 3 ppm. In the second set of experiments (Nodelman and Ultman 1999b), healthy nonsmoking subjects were administered 3 ppm chlorine gas boluses at respiratory flows of 150, 250, and 1000 mL/sec, and it was determined that under the conditions of their experiments more than 95% of the inspired chlorine was absorbed in the upper airways, whereas less than 5% of inspired chlorine penetrated beyond the upper airways and none reached the respiratory air spaces.

The pathophysiology of chlorine exposure results from the strong oxidizing nature of chlorine. Chlorine forms both hypochlorous and hydrochloric acid on contact with moist mucous membranes. Due to the high water content of epithelial lining fluid in the upper airways, chlorine hydrolysis occurs rapidly and with such a large equilibrium constant that the concentration of chlorine in the form of the acid ions is approximately 120,000 times the concentration of molecular chlorine. That is, the effective solubility of chlorine between respired gas and the mucous phase is five orders of magnitude greater than the physical solubility (Nodelman and Ultman 1999a). Hypochlorous acid decomposes into hypochloric acid and free oxygen radicals ($O_2^-$), which may disrupt the integrity and increase the permeability of the epithelium (WHO 1982) and, by an increase in hydrogen ions, decrease the blood pH, provided that a sufficient dose of chlorine gas is absorbed (Wood et al. 1987). Chlorine may also react with the sulfhydryl groups of amino acids, thereby inhibiting various enzymes (WHO 1982; McNulty et al. 1983). Tissue damage is a result of the disruption of cellular proteins (Ellenhorn and Barceloux 1988).

*Sensory Evaluation.* Chlorine has a readily identifiable "bleachy" odor. However, determinations of the order threshold for chlorine have given varying results. The odor threshold of chlorine has been reported to range between 0.2 and 3.5 ppm (Ryazanov 1962; Fieldner et al. 1921; Leonardos et al. 1968), although there has been a general agreement that the threshold of odor perception is between 0.2 and 0.4 ppm, with considerable variation among subjects. The initial detection odor threshold of chlorine, that is, when the odor of chlorine is first detected, is 0.06 ppm (Stokinger 1981).

In a review of odor thresholds and nasal irritation levels of several chemical substances, Ruth (1986) listed the odor threshold for chlorine to range from a low of 0.01 ppm (reported as 0.03 mg/m$^3$) to a high of 5 ppm (reported as 15 mg/m$^3$), with a bleachy, pungent odor. The irritating concentration was listed as 3 ppm (reported as 9 mg/m$^3$). In a compilation of odor thresholds of 214 industrial chemicals with Threshold Limit Values (TLVs), chlorine was listed as having an air odor threshold (geometric average) of $0.31 \pm 1.8$ ppm (SE) (Amoore and Hautala 1983). Amoore and Hautala (1983) also calculated an odor safety factor of 3.2 for chlorine. An odor safety factor is the threshold limit value (for chlorine, 1 ppm) divided by the odor threshold (0.31 ppm). This odor safety factor falls into an odor safety classification of C; less than 50% of distracted persons perceive the odor warning of the TLV.

Leonardos et al. (1968) evaluated the effect of chlorine on the sense of smell under controlled laboratory conditions. Using a trained odor panel, chlorine was presented to the panel in a static air system using low-odor background air as the dilution medium. The odor threshold was defined as the first concentration at which all four panel members could detect the odor. The odor threshold under these conditions was reported to be 0.314 ppm. Ryazanov (1962) reported that the odor threshold for chlorine ranged from 0.3 to 0.4 ppm (reported as 0.8–1.3 mg/m$^3$).

Beck (1959) and Rupp and Henschler (1967) exposed human volunteers in exposure chamber experiments to varying concentrations of chlorine and for varying lengths of time to measure odor detection and perception. Beck (1959) exposed 10 healthy subjects to chlorine at concentrations ranging from 0 to 1.0 ppm. At 0.09 ppm, 7 of 10 subjects recognized the odor as chlorine and 4 of 10 noted a very slight respiratory irritation as a slight tickling sensation; at 0.2 ppm, all noticed the odor and experienced slight respiratory irritation; and at 1 ppm, 7 of 10 experienced annoying symptoms (nose, throat, and eye irritation). The exposure was terminated after 20 min. In addition, he noted a decrease in odor perception by the subjects. To further evaluate the observation of a decrease in odor perception, Beck (1959) exposed an additional 10 subjects to chlorine gas at 0.44 ppm, and 4 perceived the odor; however, after 1–24 min, odor perception decreased to a point at which it was no longer detected. As the concentration increased, the number of subjects detecting the odor increased and the length of time for which the odor was detected also increased. Rupp and Henschler (1967) exposed 20 subjects to chlorine gas for a period of 30 min. Seven of 14 persons detected the smell of chlorine at an average concentration of 0.02 ppm, all 20 subjects detected the odor at an average concentration of 0.45 ppm, and at concentrations averaging 0.72 ppm, all subjects correctly recognized the odor as chlorine.

In a separate set of experiments (Rupp and Henschler 1967), chlorine was slowly introduced to the exposure chamber, which the subjects had already entered. The odor of chlorine was first detected as 0.06 ppm, and by 0.2 ppm all subjects exposed could smell the gas. When comparing these two studies, it is likely that in the first set of exposures the concentration of chlorine in the chamber dropped considerably when the subjects entered, thus skewing the results. Odor perception was also evaluated in this second set of exposures. Similar to the results reported by Beck (1959), it was found that the ability to perceive chlorine was not constant. There was a positive correlation between the length of time of odor perception and the chlorine concentration; i.e., the lower the concentration, the shorter the length of time of perception, the higher the concentration, the longer that the subjects were able to perceive the odor. In conclusion, a relatively rapid tolerance develops to chlorine at low concentrations. However, at higher concentrations, subjects continue to perceive the odor.

*Acute Toxicity.*    Chlorine gas was used by the Germans as a chemical warfare agent in 1915 (Sandall 1922; Penington 1954; Anonymous 1984). The symptoms as described by field doctors included "irritable heart," shortness of breath on exertion, bronchitis, and pseudotuberculosis with symptoms of low fever, profuse expectoration, loss of weight, and pulmonary symptoms that were suggestive of a tuberculous lesion (Meakins 1919). The clinical presentation of acute chlorine gas poisoning is dependent on concentration of chlorine and duration of exposure as well as water content of tissue involved and presence of underlying cardiopulmonary disease (Ellenhorn and Barceloux 1988). In general, clinical signs observed in acute animal studies are similar to those seen in acute accidental exposures in humans. In mild exposures, clinical signs include lacrimation, conjunctival irritation, rhinorrhea, cough, headache, sore throat, chest pain, dyspnea, nausea, and pulmonary function deficits. After more severe exposures, clinical signs include ulcerative tracheobronchitis, pulmonary edema, respiratory failure, and death. Corneal abrasions and cutaneous burns from direct exposure can also occur (Ellenhorn and Barceloux 1988). Ellenhorn and Barceloux (1988) compiled a listing of chlorine exposure thresholds and estimated clinical effects:

• 0.2–3.5 ppm: odor detection
• 1–3 ppm: mild, mucous membrane irritation, tolerated up to 1 hr
• 5–15 ppm: moderate irritation of the respiratory tract
• 30 ppm: immediate chest pain, vomiting, dyspnea, cough
• 40–60 ppm: toxic pneumonitis and pulmonary edema
• 430 ppm: lethal over 30 min
• 1000 ppm: fatal within a few minutes

Information regarding the acute toxicity of chlorine gas exposure to man has been acquired either through human volunteer experiments in which subjects are exposed to low levels of chlorine gas for short time periods or through clinical reports from chlorine gas poisonings.

*Experimental Exposures.*    Anglen (1981) exposed 30 volunteer college students to 0, 0.5, 1, or 2 ppm for either a 4-hr or an 8-hr exposure period. The subjects recorded their reactions and perceptions (smell, taste, itching or burning of the eyes, nose, throat, production of tears, dizziness, drowsiness, and shortness of breath) at scheduled time intervals during the exposure period. Eye irritation was evaluated by a comparison of close-up photographs taken of the left eye before exposure and 2, 4, and 8 hr into the exposure. The photographs were evaluated by an ophthalmologist and graded for degree of change. In addition, pulmonary function tests, including forced vital capacity (FVC) and forced expiratory volume in 1 sec ($FEV_{1\,sec}$) were measured preexposure, and 2, 4, and 8 hr into the exposure. The results indicated that odor was not a good indicator of exposure. Itching and burning of the throat had the highest response near the end of the 8-hr exposure to 1 ppm chlorine. Responses for sensations of itching or burning of the nose and eyes and general discomfort were significantly greater than during the control exposures, mostly at 1 ppm chlorine. Few positive responses were recorded for nausea, headache, dizziness, and drowsiness; the photographs showed no evidence of eye irritation and several subjects reported shortness of breath. In addition, there were some significant gender-related differences in the responses. Overall, the indices of irritation were higher for males than for females and a higher percentage of females than males responded positively to the sensation of headache and drowsiness and general discomfort. Exposure to 1 ppm or higher concentrations of chlorine for 8 hr produced significant changes in pulmonary function and increased subjective irritation. At 1.5 ppm chlorine, 6 of 14 subjects showed increased mucous secretion from the nose and increased mucus in the hypopharynx.

Rotman et al. (1983) demonstrated that exposure to 1 ppm chlorine for 8 hr produced significant, although transient, changes of pulmonary function in healthy volunteers. Volunteers were exposed for two 4-hr periods to either 0.5 ppm or 1 ppm chlorine gas. Each volunteer was subjected to pulmonary function tests preexposure and 2 and 24 hr postexposure. While in the exposure chamber, the volunteers were subjected to 15 min of exercise/hr. Exposure to 0.5 ppm resulted in no subjective symptom reporting by the volunteers, and only trivial changes in the pulmonary function tests following the exposures. Those in the 1-ppm group reported itchy eyes, runny nose, and a mild burning in the throat. Significant differences were seen in FVC, $FEV_1$, peak expiratory flow rate, forced expiratory flow rate at 25% and 50%, and airway resistance as compared to baseline.

More recently, D'Alessandro et al. (1996) evaluated the exaggerated response to chlorine inhalation among persons with nonspecific airway hyperreactivity as compared to normal subjects. A total of 15 subjects were evaluated, 10 with and 5 without airway hyperresponsiveness (HR). In the first group of exposures, 10 subjects were evaluated, 5 with and 5 without HR after a 60-min exposure to 1 ppm chlorine. In the second group of exposures, five subjects with HR were exposed for 60 min to 0.4 ppm chlorine. The 0.4-ppm concentration was chosen based on the 0.5 ppm recommended ACGIH TLV-TWA for chlorine. The 1-ppm concentration of chlorine was chosen based on the ACGIH recommended occupational 15-min ceiling short-term exposure limit (STEL). Airflow and airway resistance were measured immediately before and after exposure. In addition, at 24 hr before and 24 hr after exposure, lung volumes, airflow, diffusing capacity, airway resistance, and responsiveness to methacholine were evaluated. After exposure to 1 ppm chlorine, there was a significant decrease in $FEV_1$

among both groups; however, the HR group demonstrated an approximately threefold greater decrease than the normal group. In addition, specific airway resistance increased to a greater degree among the HR group as compared with normal subjects. At the 24-hr follow-up, there were no significant chlorine related deficits in either group. The exposure to 0.4 ppm did not produce any pulmonary function deficits among the HR subjects, indicating that TLV-TWA is protective of sensitive individuals. However, this work did demonstrate that at the 1.0-ppm STEL for chlorine there is a susceptible subpopulation of persons who may experience an exaggerated airway response to chlorine.

*Accidental Exposures.* Numerous cases of acute chlorine inhalation incidents have been documented in the literature, some of which have resulted in the deaths of individuals exposed to chlorine. Accidental chlorine inhalation has occurred in home, school, work, and recreational situations. Additionally, there are documented cases of voluntary chlorine inhalation as a supposed form of chemical abuse (Rafferty 1980; Dewhirst 1981). Exposures have been reported as the result of industrial accidents where the chlorine cloud migrates from an industrial site to a residence (Adelson and Kaufman 1971; Givan et al. 1989; Chester et al. 1977), leakage of chlorine canisters stored at the residential site (Levy et al. 1986), or the mixture of household chemicals that results in the release of chlorine gas (e.g., bleach and vinegar) (Malone and Warin 1945; Bloomfield 1959; Faigel 1964; Jones 1972; Warrack 1978; Gapany-Gapanavicius et al. 1982; Szerlip and Singer 1984; Phillip et al. 1985; Boulet 1988; Anonymous 1991; Mrvos et al. 1993; Deschamps et al. 1994).

Accidental exposure to chlorine gas has also been documented at swimming pools. Numerous accidental exposures to high chlorine gas concentrations have been described, either through exposure to vapors generated from pool chlorinator tablets (Wood et al. 1987; Martinez and Long 1995; Heidemann and Goetting 1991) or from chlorine gas cylinders used in the chlorination system (Decker and Koch 1978; Fleta et al. 1986; Polysongsang et al. 1982; Mustchin and Pickering 1979; Vinsel 1990).

Industrial accidents resulting in the exposure of individuals to high concentrations of chlorine have occurred on several occasions. These accidents have provided investigators with the opportunity to evaluate the long-term effects of such exposures on workers following the industrial incident. Charan et al. (1985) described the effects of accidental chlorine exposure on pulmonary function tests. The chlorine exposure occurred when a flexible hose connecting a railroad car filled with chlorine gas disconnected. It was estimated from the color of the yellow green cloud that the chlorine concentration was approximately 1000 ppm; however, no measurements of chlorine concentrations were made. No further information was provided as to the concentrations of chlorine present at the accident site. Nineteen workers were exposed from a few seconds to a few minutes. The major clinical signs were irritation of eyes, nose, throat, paroxysms of cough, tightness in the chest, and airflow obstruction due to acute inflammation in the airways. Pulmonary function tests were conducted on days 10, 40,

310, and 700 following the exposure. At the 310th day measurement, 38% (5 of 13 evaluated) still showed airway obstruction and by 700 d postexposure, 27% (3 of 11 evaluated) had evidence of pulmonary function deficits.

Schwartz et al. (1990) conducted follow-up tests on 20 subjects exposed to chlorine gas during an accidental release of liquid chlorine at a pulp mill. Pulmonary tests were performed on several occasions over 12 yr. Standard spirometry tests, methacholine challenges, and statistics were performed at 1, 10, and 40 d and 1, 2, 9, and 12 yr after exposure. On average, each subject was followed for 8.5 yr, and 13 of the exposed persons were followed for the full 12 yr. One day after exposure, there was evidence of airflow obstruction and air trapping. Over the 12-yr test period, there was a persistence of airflow obstruction, but air trapping was not highly prevalent. A progressive decline in residual volume was noted in 67% of those tested at year 12, having a residual volume below 80% of their predicted value. It was not clear whether the chlorine exposure caused the increased airflow obstruction because 70% of the workers exposed were active smokers, which may account for the observed airflow obstruction. Five of the 13 workers tested 12 yr after exposure had increased airway reactivity to methacholine challenge.

Reactive airway disease (RAD) has been associated with chlorine exposure (Boulet 1988; Donnelly and Fitzgerald 1990; Bherer et al. 1994; Lemiere et al. 1997). A diagnosis of RADS (as defined by Boulet 1988) is made when there are the following conditions:

1. Absence of previous respiratory symptoms or disease.
2. Possibility of patient to associate onset of RAD with a specific event.
3. High-level exposure to gas, smoke, fume, or vapor.
4. Onset of symptoms recurring within a few hours of exposure.
5. Asthma-like symptoms such as cough, wheezing, and dyspnea.
6. Pulmonary function tests show airflow obstruction and increased methacholine response.
7. No evidence of other pulmonary disease.

Donnelly and Fitzgerald (1990), Bherer et al. (1994), and Lemiere et al. (1997) described case studies in which individuals were exposed to a "high" concentration of chlorine gas. The lung injuries after each case of chlorine gas inhalation were well recognized. Clinical symptoms after the chlorine gas exposure included immediate eye, nasal, and throat burning, retrosternal burning, wheezing, acute pulmonary edema, laryngotracheobronchitis, obstructive ventilatory impairment, and bronchitis obliterans. Bherer et al. (1994) studied the prevalence of persistent respiratory symptoms and bronchial hyperresponsiveness due to RAD in a population of construction workers exposed to chlorine gas during the renovation of a pulp and paper mill. Of subjects at a moderate to high risk of developing RAD after the chlorine exposure, 82% still had respiratory symptoms, 23% showed evidence of bronchial obstruction, and 41% demonstrated bronchial hyperresponsiveness 18 to 24 mon after exposure. Donnelly and Fitzgerald (1990) described a case study in which a 30-yr-old male worker was sprayed in the face with chlorine gas. More than 6 yr after the chlorine gas exposure, the symptoms of nocturnal or exercise induced wheezing persisted despite inhalant therapies.

Catastrophic community exposures to chlorine gas most commonly result from storage or transportation accidents involving the pressurized liquid form of chlorine. In 1961, approximately 6000 gal (36 t) of liquid chlorine was spilled in La Barre, LA, as a result of a train accident (Joyner and Durel 1962). The chlorine gas cloud covered a 6-square-mile area and approximately 1000 people were evacuated. Three hours after the spill, the concentration of chlorine at the outer fringes of the contaminated area (approximately 180 m away) was 10 ppm. Seven hours after the spill, measurements of 400 ppm chlorine were recorded 60 m away from the spill; however, this concentration was not thought to represent the maximum concentration. Concentrations of 8 ppm were recorded 9.5 hr after the initial release at the site of the wreck. One hundred people were exposed to the chlorine gas. All heavily exposed victims expressed severe dyspnea, coughing, vomiting, frothy sputum, and rales. Most complained of burning eyes and had acute conjunctival injection with profuse tearing and photophobia. Some developed first-degree burns on the skin from the vapors. Of those exposed, 10 developed pulmonary edema and 1 person died. Chest X-rays on the 3rd and 4th days post exposure showed fine mottling distributed bilaterally and symmetrically in both lungs, which cleared by the 10th and 12th day post exposure. All patients were released 16 da post exposure.

Similar events have been described by Chasis et al. (1947), Kaufman and Burkons (1971), Colardyn et al. (1976), Cordasco et al. (1977), Jones et al. (1986), and Achon and Roberts (1996). Schroff et al. (1988) evaluated the respiratory cytopathology of victims of a chlorine gas accident in Bombay, India. At a distance of 17–20 m from the leak, chlorine concentrations were measured at 66 ppm. Eighty-eight patients were admitted to the local hospital; all experienced immediate dyspnea and coughing, irritation of the throat and eyes, headache, giddiness, chest pain, and abdominal discomfort. Bronchial brushings were obtained from 28 patients on days 5, 15, and 25 postexposure. Observations on day 5 postexposure included basal and goblet cell hyperplasia, acute inflammation, and chromatolysis of the columnar epithelial cells. By days 15 and 25 postexposure, signs of epithelial regeneration and repair by fibrosis were observed in 7 patients.

*Chronic Toxicity.* The general public may potentially be exposed to chlorine through air inhalation, water ingestion, swimming pool attendance, contact with chemicals used in the household, and ingestion of chlorine processed food products (e.g., bleached flour). Despite the high prevalence of potential exposure to chlorine, little is known about its long-term inhalation effects in humans (Barbone et al. 1992). In addition, employees of facilities that use chlorine in their day-to-day operations may be exposed to low levels of chlorine. Ferris et al. (1967) compared 147 men who worked in a pulp mill and were exposed to chlorine and sulfur dioxide to 124 men from a paper mill (non-chlorine exposed) to evaluate the prevalence of chronic respiratory disease in these facilities. Air sampling was conducted for sulfur dioxide and chlorine in both facilities. A survey of occupational and smoking history was conducted and a standard respiratory questionnaire completed by the study participants. Simple tests of pulmonary function were also conducted on the workers. No significant differences were found in respiratory symptoms or in simple tests of ventilatory function, but men working with chlorine had a somewhat poorer respiratory function and more shortness of breath than those working with sulfur dioxide in pulp mill.

Ferris et al. (1979) conducted a 10-yr follow-up study on pulp (chlorine-exposed) and paper mill (non-chlorine-exposed) workers. A cohort of 271 men was studied in 1963 and followed up in 1973. In 1963, a subpopulation of older men who had been continuously exposed to chlorine showed slightly lower expiratory flow rates than those non-chlorine exposed. Ten years later, there were no differences in respiratory symptoms or prevalence of chronic, nonspecific respiratory disease in the chlorine-exposed workers.

In 1969, the relationship between chronic chlorine gas inhalation and obstructive pulmonary disease was studied in 139 workers in a chlorine plant (Chester et al. 1969). The average chlorine concentration in the air was less than 1 ppm, but accidental exposures occurred in 55 of the 139 workers requiring $O_2$ therapy. Of the 139 workers at risk, only 3 workers (smokers) had significant impairment of ventilatory function. However, the radiographic and respiratory questionnaires correlated poorly with the observation of abnormal ventilatory function in these 3 workers. Therefore, it was unresolved whether chlorine inhalation produces chronic pulmonary damage. Patil et al. (1970) found no difference in ventilatory function, frequency of colds, shortness of breath, chest pain, abnormal chest X-rays, or ECGs among diaphragm cell workers exposed to 0.006–1.42 ppm (few over 1 ppm) for an average period of 10.9 yr. Exposed workers did exhibit increased anxiety, dizziness and fatigue, and tooth decay in those chronically exposed above 0.5 ppm as compared to nonexposed workers. Although these increases were significant when compared to control, they were based upon self-reported subjective symptoms, and no real dose response was seen. Other environmental factors unique to diaphragm cell rooms may have played a role in these symptoms instead of chlorine exposure.

*Carcinogenicity.*   There is no evidence of carcinogenic effects resulting from chlorine inhalation exposure in humans. Epidemiological studies have evaluated the relationships between chlorine exposure and brain tumors (Bond et al. 1983), chlorine and renal tumors (Bond et al. 1985), and chlorine and lung cancer (Bond et al. 1986; Jäppinen et al. 1987; Heldaas et al. 1989; Barregård et al. 1990; Barbone et al. 1992).

No association was found for chlorine and brain tumors (Bond et al. 1983). An increased odds ratio for renal cancer in a chlorine production area was found during an in-plant case control study of 26 renal cancer deaths (Bond et al. 1985). However, this increase in cancer incidence was not attributed to chlorine because there was a lack of increase in renal cancer incidence in the magnesium processing area, which uses large amounts of chlorine. The increased incidence was attributed to asbestos and caustics used in the area.

Jäppinen et al. (1987) evaluated the cancer incidence of workers in the Finnish pulp and paper industry who had a potential risk for chlorine exposure. Lung cancer excess was observed in 78 men versus 62.6 expected. The lung cancer excess was most prominent in board mill workers after 20 yr of latency. The relative risk (RR, the ratio of observed incidence versus expected incidence) for lung cancer was 3.2 (confidence interval, 95% CI: 2.1–4.8) and was not explained by cigarette smoking. However, this finding has not been supported in studies of other exposed groups (Bond et al. 1986; Heldaas et al. 1989; Barregård et al. 1990; Barbone et al. 1992).

Bond et al. (1986) conducted a nested case-control study of lung cancer workers at a major Texas chemical production facility, a mortality cohort epidemiological study of 19,608 employees suggested an increased risk of lung cancer compared to the U.S. Between 1940 and 1981, 308 deaths were attributed to lung cancer; however, no association was found with chlorine exposure and the increase in lung cancer at the facility. The odds ratio for developing lung cancer following chlorine exposure was 1.08 (confidence interval, 0.81–1.44).

In a Norwegian cohort study of magnesium production workers with a potential for chlorine exposure, the RR for lung cancer was 1.8 (95% CI: 1.2–2.5), and the excess of cancer incidence was evident among workers with 10 yr or more in the chlorination areas and with at least 20 yr since first employment. In a subset of workers who experienced chlorine gas intoxication, 4 lung cancer cases occurred versus 1.3 expected among this group (RR = 3) (Heldaas et al. 1989). In a Swedish study of chlor alkali workers (Barregård et al. 1990) exposed to chlorine, the RR for lung cancer was 2.0 (95% CI: 1.0–3.8). However, neither of these studies was able to adequately control for confounding factors including smoking or other occupational exposures, such as asbestos. Similarly, an increased incidence of lung cancer was noted in a case control study evaluating the relationship between lung cancer and occupational factors among employees at a dye and resin manufacturing plant (Barbone et al. 1992). An elevated odds ratio for lung cancer was observed for employees who were seen at the plant for acute exposure to chlorine. However, the observed association between lung cancer and acute chlorine exposure was based upon small numbers of workers and thus may be the result of chance or confounding by other unidentified exposures not attributed to chlorine exposure.

*Teratogenicity and Reproductive Toxicity.*    There is no evidence of teratogenic or reproductive toxicity caused by chlorine exposure in humans. Sklyanskaya and Rappoport (1935) evaluated the outcome of 15 pregnancies among female workers at a chlorine plant in 1932–1933. The authors concluded that pregnancy, delivery, pueperium, and lactation were not affected.

*Neurotoxicity.*    There have been rare reports in the literature associating neurotoxicity with chlorine gas exposure (Adelson and Kaufman 1971; Levy et al.

1986; Kilburn 1995; Hirsch 1995). Adelson and Kaufman (1971) were the first to suggest a possible connection in a case study in which a young woman was exposed to chlorine gas while sleeping during a release at a nearby wastewater treatment plant. The woman exhibited confusion, became lethargic and comatose 57 hr after exposure, and died within 72 hr after the exposure. On autopsy, the brain was swollen with marked convolutional flattening and foci of recent focal and confluent subarachnoid hemorrhages and areas of parenchymal hemorrhage. These brain lesions had not been previously described in victims of fatal chlorine poisoning. The authors hypothesized that the lesions were caused by hypoxia resultant from the widespread pulmonary changes. The woman's husband, who was also exposed, died after 25 hr with no effects on the brain.

Levy et al. (1986) described cerebral lesions evident in a woman exposed to chlorine gas as a result of mixing household chemicals, although no relationship between the lesions and the exposure was established. The woman had symptoms of headache, vomiting, ataxia, and the inability to concentrate. Over the next 2 yr the neurological symptoms slowly improved, although hypersensitivity to cold and some cognitive disability persisted. A magnetic resonance imaging scan showed multiple areas of decreased signal in the periventricular white matter.

In connection with litigation, Kilburn (1995) and Hirsch (1995) have both evaluated a few of the thousands of people exposed to a large fugitive release of chlorine that occurred in Nevada. The information in the papers was too sparse to determine any strength of association between the reported respiratory and neuropsychological problems and exposure to chlorine. There were no chlorine exposure data for the six subjects reporting neurological symptoms. When examined by Kilburn 15–50 mon after exposure, the six Nevada subjects exhibited various complaints and impairments, including memory loss, excessive fatigue, tinnitus, dizziness and loss of balance, loss of strength, difficulty sleeping, depression, and irritation. Kilburn administered neurophysiological tests such as simple reaction time, body balance, blink reflex, color discrimination, visual fields, hearing acuity, and neuropsychological tests and found disturbances of blink, balance, hearing, reaction time, vision, and verbal and visual recall that he speculated could have been caused by exposure to an unknown quantity of chlorine. The control group that Kilburn used was from an earlier study and was not matched by age and sex (Kilburn deposition, October 4, 1995, Horne v. Pioneer ChlorAlkali Company, Inc., U.S. District Court, District of Nevada).

Kilburn (1995) reached the same conclusion with respect to one patient who was sprayed in the face with chlorine at a wastewater treatment facility. Hirsch provided a brief abstract (Hirsch 1995) describing four of the Nevada subjects with persistent sequelae following medical, psychiatric, neurological, neuropsychological, chemosensory, and psychometric testing several months after exposure, Hirsch provided a clinical diagnosis of neurotoxicity that he attributed to exposure to unknown quantities of chlorine.

Because of study limitations, there has been no clear correlation between chlorine exposure and the effects attributed by Kilburn and Hirsch to chlorine.

## B. Animal Exposures: Inhalation

The toxicity of inhaled chlorine varies greatly as reported in the literature. This variability may be related to species differences, exposure chamber differences, exposure method differences (whole body versus nose only), variations in atmosphere generation, analytic determination of the chlorine concentration in the chamber, and the sophistication of the study itself. This variability is not surprising considering that these observations regarding the toxicity of inhaled chlorine have been made over a time span of greater then 50 yr, during which experimental procedures have been greatly modified.

*Pharmacokinetics.* There is no published literature regarding the absorption, distribution, metabolism, and excretion of molecular chlorine following inhalation exposure.

*Sensory Evaluation.* Sensory irritation and pulmonary function studies have been conducted in rats, mice, and rabbits after exposure to chlorine gas. Sensory irritation of an airborne chemical can be quantified by measuring decreases in respiratory rate during the inhalation exposure to different concentrations of the irritant (Alarie 1973). Sensory irritation in rodents is quantified using the concentration expected to produce a 50% decrease in respiratory rate ($RD_{50}$). Barrow et al. (1977) exposed mice for 10 min to concentrations of chlorine ranging from 0.7 to 38.4 ppm and hydrogen chloride from 40 to 943 ppm. The no observed adverse effect level (NOAEL) for the sensory irritation endpoint for chlorine and hydrogen chloride was 0.7 and 40 ppm, respectively. The $RD_{50}$ for chlorine and hydrogen chloride were 9.3 and 309 ppm, respectively. These studies showed that chlorine was 33 times more potent than hydrogen chloride as a sensory irritant.

*Acute Toxicity.* Effects seen in experimental animals as a result of acute inhalation exposure to chlorine include blinking of the eyes, sneezing, lacrimation, coughing, inflammation of the conjunctiva, and labored breathing with cyanosis and pain. With increasing concentrations, effects include injury to the respiratory tract tissue, bronchial spasm, delayed pulmonary edema, and immediate or delayed death. The lethality of chlorine in rodents and dogs has been evaluated (Underhill 1920; Silver and McGrath 1942; Schlagbauer and Henschler 1967; Vernot et al. 1977; Back et al. 1972; Zwart and Woutersen 1988). Table 5 summarizes the acute lethality data for mice, rats, and dogs. Based upon its 1-hr $LC_{50}$, chlorine appears to be a serious to moderate hazard.

The 50% lethal concentration ($LC_{50}$) is that concentration which will kill half the animals in the allotted time frame and is a measure of acute toxicity. In general, 1-hr $LC_{50}$s $\leq 20$ ppm are considered severe hazards, >20 to 200 ppm are serious hazards, >200 to 2,000 ppm are considered moderate hazards, >2000 to 10,000 ppm are considered slight hazards, and >10,000 ppm are considered minimal hazards, based upon the National Paint & Coatings Association Hazardous Materials Identification System (HMIS) rating system (NAPIM 1984).

Table 5. Summary of acute lethality data.

| Species | Time | LC$_{50}$ | Reference |
|---|---|---|---|
| Mouse | 10 min | 618 ppm | Silver and McGarth 1942 |
| | 30 min | 127 ppm | Schlagbauer and Henschler 1967 |
| | 30 min | ~496 ppm[a] | Zwart and Wouterson 1988 |
| | 1 hr | 137 ppm | Back et al. 1972; Vernot et al. 1977 |
| Rat | 30 min | ~689 ppm[b] | Zwart and Wouterson 1988 |
| | 1 hr | ~449 ppm[c] | Zwart and Wouterson 1988 |
| | 1 hr | 293 ppm | Back et al. 1972; Vernot et al. 1977 |
| Dog | 30 min | 650 ppm | Underhill 1920 |

[a]Reported as 1462 mg/m$^3$.
[b]Reported as 2033 mg/m$^3$.
[c]Reported as 1321 mg/m$^3$.

Winternitz et al. (1920) conducted an extensive pathological study evaluating the pulmonary damaging effects observed postmortem in 326 dogs exposed to lethal or near-lethal concentrations of chlorine (Underhill 1920). Briefly, animals that died within 2 d of exposure evidenced severe injury to the mucous membranes of the upper respiratory tract with irregular dilation and contraction of the bronchi, resulting in alternating patches of acute emphysema and atelectasis in the lungs. All tissues showed extreme congestion and edema. An acute inflammatory reaction developed within a few hours. Animals that died 2–5 d after exposure evidenced an increased intensity of inflammation and development of lobular pneumonia, frequently complicated by abscess formation and gangrene. Bronchiolar spasm was very pronounced. Animals that died 5–15 d after exposure displayed pulmonary damage of severity between that of the animals dying within 2–5 d and the surviving animals. Death was usually due to infection, pneumonia, or bronchitis. The surviving animals displayed marked emphysema associated with bronchitis obliterans. The lung tissue surrounding the bronchus showed organizing pneumonia, and the alveoli were filled with cellular exudate. The authors hypothesized that the patchy distribution of pulmonary damage as seen in these animals represented the effects of chlorine that reached the terminal alveolar sacs through airways which were not occluded by bronchospasm.

Barrow and Smith (1975) conducted studies examining the pulmonary function changes in rabbits exposed to chlorine gas. Four rabbits per group were exposed to chlorine at concentrations of 0, 50, 100, or 200 ppm for 30 min. One animal per group was killed 30 min after the termination of exposure and one per group was killed after 3, 14, and 60 d. A NOAEL of 50 ppm was found for lung function and gross pathological changes. At higher doses, a frothy exudate in the trachea and hemorrhagic surface of the lung was observed during the postmortem examination. A decrease in pulmonary compliance was noted. The pulmonary function changes were linked with the degree of pulmonary edema and severity of anatomical emphysema and distortion of the alveolar structure and elasticity. The lung anomalies showed a correlation with the magnitude of chlorine concentration and the length of time allowed for recovery post exposure.

*Subacute Toxicity.*   Subacute inhalation exposures (multiple high dose exposures) of animals to chlorine gas result in acute pulmonary damage (Elmes and Bell 1963; Bell and Elmes 1965), lesions in the nasal cavity (Jiang et al. 1983), and development of tolerance to the sensory irritant properties of chlorine (Barrow and Steinhagen 1982; Chang and Barrow 1984). Elmes and Bell (1963) observed that rats with spontaneous pulmonary disease (SPD) had higher mortality when exposed to 16 ppm chlorine for 1 hr/d for 4 wk or when exposed to 40 ppm for 2 hr/d for 5 wk.

Rats with SPD developed acute necrosis of the bronchial mucosa and acute inflammation with cellular exudate into the periphery. Bell and Elmes (1965) determined that specific-pathogen-free rats (SPF) tolerated higher doses of chlorine than diseased animals. Young SPF rats were exposed to 90 ppm chlorine for 3 hr/d for 20 d or 104 ppm for 3 hr/d for 6 d. SPD rats were more sensitive to chlorine inhalation than SPF rats based on mortality. SPF animals that died showed similar pathological findings as SPD animals; however, the SPD rats exhibited a more severe cellular response. Schlagbauer and Henschler (1967) reported that mice exposed to 2.5 and 5.0 ppm for 8 hr/d for 3 consecutive days showed clinical signs of toxicity, reported as body weight loss. Microscopic examination of the lungs of the 5.0-ppm-exposed mice showed similar patterns of inflammation as seen in lethal or near-lethal short-term exposures.

Subacute exposure studies of rats and mice to the $RD_{50}$ concentrations of chlorine have been conducted to determine if tolerance to sensory irritation develops. Barrow and Steinhagen (1982) determined the $RD_{50}$ of chlorine in Fisher-344 rats to be 25 ppm. When groups of animals were pretreated with chlorine at concentrations of 0, 1, 5, or 10 ppm for 6 hr/d, 5 d/wk for 2 wk, a concentration-dependent increase in $RD_{50}$ values was noted, indicating a development of tolerance to the sensory irritant effects of chlorine. The $RD_{50}$ of rats pretreated with 1, 5, or 10 ppm chlorine were 90, 71, and 454 ppm, respectively. Similar results were found by Chang and Barrow (1984). Those animals preexposed to chlorine evidenced tolerance to the sensory irritant effects of chlorine gas. Animals preexposed to 10 ppm chlorine for 4 or 10 d evidenced the greatest increase in $RD_{50}$, while those exposed to 2.5 ppm evidenced some tolerance but only when preexposed for 10 d. A decrease in tolerance was seen when rats were exposed to 10 ppm for 4 d and allowed to recover for 7 d. However, the tolerance was still significant as compared to controls. The development of tolerance during exposure to chlorine is important because it indicates that irritation may be diminished with repeated exposures.

Jiang et al. (1983) evaluated the pathology in the nasal passages of rats and mice associated with the $RD_{50}$ concentration of chlorine gas. Mice and rats were exposed to 0 to 9 ppm chlorine, 6 hr/d for 1, 3, or 5 d. Lesions were observed in both the olfactory and respiratory epithelia. The lesions were most severe on the free margins of the naso- and maxilloturbinates and the anterior aspect of the ethmoid turbinates. Pathological changes included acute epithelial degeneration with epithelial cell exfoliation, erosion, and ulceration after 1–3 d exposure and squamous metaplasia after 5 d of exposure. Buckley et al. (1984) conducted similar studies in mice. Mice were exposed to 9.34 ppm for 6 hr/d for 5 d. Chlorine exposure induced moderate to severe terminal bronchiolitis in the lower respiratory tract, with occlusion of the affected bronchioles by serocellular exudate. The lesions in the nasal cavity showed a distinct anterior-to-posterior severity gradient. It was hypothesized that higher concentrations of chlorine damage the deep lung because of the low water solubility of chlorine.

Dodd et al. (1980) exposed rats to 12 ppm chlorine for 6 hr/d for 1, 5, or 10 d and evaluated the effect of exposure on lung protein or nonprotein sulfhydryl levels. No increase was seen immediately following exposure; however, an increase was seen during the 6-d recovery period of those animals exposed to chlorine for 10 d. This increase was attributed to tissue repair mechanisms. McNulty et al. (1983) exposed male Fisher-344 rats to 2.5, 5, or 10 ppm chlorine for either 1, 3, 6, or 12 hr and exposed a second group to 2.5 ppm for 6 hr/d for 5 d. Respiratory and olfactory and nasal mucosal tissues were evaluated for total sulfhydryl content (TSH). Oxidation of TSH was observed in respiratory but not olfactory mucosa. In addition, there was no deep lung damage, although previous work (Buckley et al. 1984) demonstrated lung lesions at a comparable concentration (9.3 ppm). It was concluded that the oxidation of TSH was not a good indicator of the ability of chlorine to react in a toxic manner with tissue components (McNulty et al. 1983).

*Subchronic Toxicity.* The results of a study evaluating the subchronic inhalation exposure of rats to chlorine at concentrations of 1, 3, or 9 ppm demonstrated that chlorine produces pulmonary effects at all levels of exposure and hepatic and renal effects at exposures greater than 3 ppm (Barrow et al. 1979). Rats were exposed to 0, 1, 3, or 9 ppm chlorine for 6 hr/d, 5 d/wk for 6 wk. Clinical sings observed in groups exposed to 3 and 9 ppm included ocular and upper respiratory tract irritation, lacrimation, hyperemia of the conjunctiva, nasal discharge, salivation, and gasping.

Some irritant effects were seen at 1 ppm; however, they were occasional and slight. There was some mortality in female rats exposed to 9 ppm (3 of 10 rats died before the end of the exposure). A reduction in body weight gain was seen at all three exposure levels in females and at 3 and 9 ppm for males. Pathology of the animals exposed to 1 and 3 ppm chlorine revealed inflammation of the nasal turbinates and the submucosal epithelium of the trachea. Slight to moderate inflammation was observed in the bronchioles and alveolar ducts, with an increase in the number of alveolar macrophages in the alveoli. Those animals exposed to 9 ppm evidenced inflammation of the upper and lower respiratory tract. Inflammation of the nasal turbinates and erosions of the mucosal epithelium were observed, and inflammation and epithelial hyperplasia were seen in the trachea and bronchiolar area and in the bronchioles and alveolar ducts. In addition to the pulmonary changes, rats exposed to 9 ppm demonstrated slight degenerative changes in the renal tubules of the kidney. Animals exposed to 3 and 9 ppm also showed degenerative changes to the liver as indicated by elevated serum enzymes.

*Chronic Toxicity.*   The results of studies conducted with rats, mice, and monkeys chronically exposed to chlorine have demonstrated that chlorine gas is an upper respiratory tract toxicant. The Chemical Industry Institute of Toxicology (CIIT) completed a 2-yr inhalation toxicity study in rats and mice in 1993, the results of which were published by Wolf et al. (1995). Mice and rats were exposed (whole body) to chlorine gas at concentrations of 0, 0.4, 1.0, or 2.5 ppm for 2 yr. This exposure regimen resulted in decreased body weight gain in mice and rats; however, survival rates were not affected. Chlorine-induced lesions were found only in the nasal passages. Chlorine-induced lesions were most severe in the anterior nasal cavity and included respiratory and olfactory epithelial degeneration, septal fenestration, mucosal inflammation, respiratory epithelial hyperplasia, squamous metaplasia, goblet cell hypertrophy and hyperplasia, and an apparent increase in nasal mucous cells.

A 1-yr inhalation toxicity study of chlorine in rhesus monkeys were conducted by Klonne et al. (1987). Rhesus monkeys were exposed to concentrations of 0, 0.1, 0.5, or 2.3 ppm chlorine gas for 6 hr/d, 5 d/wk for 1 yr. Monkeys exposed to 2.4 ppm chlorine exhibited daily signs of ocular irritation and a superficial conjunctival irritation that was present at study termination. There were no treatment-related effects on body weight, on neurological or electrocardiographic parameters, or on pulmonary physiology (pulmonary diffusing capacity or distribution of ventilation). There were no treatment-related macroscopic changes in tissues of either gender in any exposure group. Treatment-induced lesions as determined histopathologically were confined to the respiratory tract. The histopathological changes were found in the respiratory epithelium of the nasal passages and trachea and were limited to focal, concentration-dependent related changes. The changes in the nose and trachea were mild, focal epithelial hyperplasia with loss of cilia and decreased number of goblet cells in the affected areas (trachea in females only; nasal in males and females). At lower chlorine concentrations, similar although less prominent respiratory lesions were seen. The authors concluded that 2.3 ppm acted as an eye and respiratory irritant in the monkeys, while 0.1 and 0.5 ppm induced changes of questionable significance. They further stated that the monkey appears to be less sensitive than the rat to chlorine.

A comparison of the histopathology of the rhesus monkey study (Klonne et al. 1987) and the CIIT mouse and rat study (Wolf et al. 1995) was conducted to evaluate the differences in histopathological findings (Leininger et al. 1994). In the CIIT study, mice and rats exposed to chlorine for 2 yr had an apparent increased number of nasal mucous cells. However, the reported response in rhesus monkeys exposed to chlorine gas for 1 yr consisted of small foci of respiratory epithelial hyperplasia with no apparent increase in nasal mucosal cells. Both studies evaluated the effect of chlorine in the concentration range 0–2.5 ppm. Leininger et al. (1994) reexamined the material from these studies to quantify mucous cells and compare the responses in these species. A confirmation of mucous cell hyperplasia in the rodent nose (but not lung) was made, with a more pronounced increase in rats. There was a wide distribution of hyperplasia within the lateral aspect of the nasal and maxilloturbinates, and the ventral part of the nasal septum most prominently affected, especially anteriorly in the nose. The monkeys showed focal decreases in mucous cells on the margins of the middle turbinates.

*Carcinogenicity.*   Most animal studies have primarily focused on acute or subacute exposures over a wide range of concentrations. With a few exceptions, most chlorine toxicity studies focused on lethal or high-exposure, short-term concentrations. A 2-yr inhalation study was conducted to determine the potential carcinogenic response of prolonged exposure of chlorine in rats and mice (Wolf et al. 1995). Mice and rats were exposed (whole body) to chlorine gas at concentrations of 0, 0.4, 1.0, or 2.5 ppm for 2 yr. Male and female mice and male rats were exposed 6 hr/d, 5 d/wk for 2 yr, while female rats were exposed on 3 alternate d/wk for 2 yr. The 3 d/wk exposure regimen of the female rats was based upon unpublished observations of a greater sensitivity of female rats to chlorine. Based upon survival analyses, it was determined that the females would not tolerate a 5 d/wk exposure regimen.

The results of this study demonstrated that chlorine gas is an upper respiratory tract toxicant but not a carcinogen in rats or mice under the conditions of this study. As previously described, the exposure regimen resulted in decreased body weight gain in mice and rats; however, survival rates were not affected. Chlorine-induced lesions were found only in the nasal passages, and these lesions were generally site specific within the nose.

*Teratogenicity and Reproductive Toxicity.*   Sklyanskaya and Rappoport (1935) exposed rabbits to chlorine concentrations of 0.6–1.57 ppm (reported as 1.7–4.4 mg/m$^3$) 5 hr/d, every other day, for 1–9 mon. A normal course of pregnancy and parturition was reported for six rabbits, with the delivery of healthy, well-developed offspring. Two rabbits evidenced macerated fetuses in the abdominal cavity, which may have been in the process of being absorbed. However, this observation is difficult to attribute to an effect of chlorine exposure because of spontaneous disease complications that arose in the study and the absence of control animals in the study.

*Effects on Immune System.*   Currently, there is no information to suggest chlorine has any adverse effects on the immune system.

### C. Animal Exposures: Oral

*Subacute Toxicity.*   Chlorine and chlorine products are often used in the food industry for disinfection of meat and seafood, of preparation areas and utensils, and for bleaching flour. There is conflicting evidence whether chlorine-bleached flour and its components are toxic to rats (Cunningham et al. 1977; Cunningham and Lawrence 1978; Fisher et al. 1983a). Cunningham et al. (1977) evaluated the toxic effects of bleached cake flour and its components in rats. Three types of flour diets were prepared, each at a chlorination level of 0.2% (2,000 ppm) and 1% (10,000 ppm). A control group was fed unbleached flour of the same type. The diets were fed to rats for 2 wk. At necropsy, the animals on diets 1 (bleached cake flour) and 3 (chlorinated wheat proteins) demonstrated statistically significant decreases in body weight and increased liver weights at both dose levels. Those animals on diet 2 (chlorinated lipid fraction) demonstrated a decrease in body weights at both dose levels. Similar results were reported by Cunningham and Lawrence (1978). It was concluded that bleached cake flour is toxic to rats at the concentrations tested (Cunningham et al. 1977; Cunningham and Lawrence 1978). Conversely, when Fisher et al. (1983a) fed rats a cake-based diet made with bleached flour at levels of 0, 1,257, and 2,506 ppm for 4 wk, there was no evidence of abnormality in appearance or behavior. Although liver weights were slightly elevated, there was no decrease in body weights at any of the dose levels. In comparing the results of the Cunningham studies (Cunningham et al. 1977; Cunningham and Lawrence 1978) and their work, Fisher et al. (1983a) criticized the results of the latter studies. Fisher et al. (1983a) suggested that the poor growth rates evidenced in the controls fed unbleached flour (Cunningham et al. 1977; Cunningham and Lawrence 1978) implied either a gross deficiency in the basal diet or that "abnormal rats" were used in the studies. In addition, Fisher et al. (1983a) also stated that the accepted method of evaluating the toxicity of bleached flour is by feeding a cake-based diet, not a flour-based diet. Fisher et al. (1983a) concluded that bleached flour is not toxic to rats.

*Subchronic Toxicity.*   In a study conducted by Kotula et al. (1987), groups of male and female rats were fed diets that contained beef treated with a chlorine solution containing 0, 50, 200, or 600 ppm free chlorine. The beef was processed, freeze-dried, and mixed with the rat diet, which was fed to rats for 92 d. The actual dose of chlorine to which the animals were exposed was not reported. Analyses included various blood parameters and enzyme activities, and urinalysis was performed on each animal. Significantly increased prothrombin clotting times of males in the 600-ppm group was the only finding. The authors considered the prothrombin results of "questionable importance" and concluded that the bleached beef diet was not associated with clinical or hematological abnormalities.

*Chronic Toxicity and Carcinogenicity.*   No direct evidence implicating chlorine or hypochlorites as carcinogens was found in the literature.

Chronic exposure to chlorine-processed food products (such as bleached flour) occurs in the human population. Work by Ginnochio et al. (1983) and Fisher et al. (1983b) evaluated the toxicity of bleached cake flour in a long-term feeding study in mice and rats. Mice were fed cake flour with chlorine levels of 0, 1,250, and 2,500 ppm for 16 or 17 mon (males and females, respectively). The most distinct observation seen in this study was a reduced survival rate. No differences in health and behavior of the mice were attributable to the bleached flour treatment. There were also no consistent treatment-related effects observed in the hematological, biochemical, and renal function studies. The study was terminated early as the mice fed the cake diet died of obesity (Ginnochio et al. 1983). The authors noted that terminating the study early limited the study because there was a lower probability of observing tumors on the shortened time scale.

Wistar rats were fed cake flour with chlorine levels of 0, 1,250, and 2,500 ppm for 104 wk (Fisher et al. 1983b). There were no differences in appearance, health, behavior, or mortality attributable to the bleached flour treatment. No carcinogenic effect was noted in either of these studies.

*Teratogenicity and Reproductive Toxicity.*   No adverse effects on fertility were noted in a seven-generation study in rats given 100 mg/L chlorine in drinking water (Druckery 1968). Smith et al. (1985) performed a study on male and female rats dosed with aqueous solutions of chlorine (1.0, 2.0, and 5.0 mg/kg). Males were dosed for 56 d and then for the 10-d breeding period. Females were dosed for 14 d before the breeding period and then for the entire breeding, gestational, and lactational periods (66 d total). Some of the pups were also dosed following weaning (21 d old), and were observed at 40 d old for reproductive toxicity. No adverse effects were noted in any of the male or female rats in this study.

Carlton et al. (1986) exposed both male and female Long-Evans rats to chlorine (0, 1, 2, or 5 mg/kg/day) by oral gavage. Males were dosed for 56 d and females for 14 da before breeding and throughout the 10-d breeding period. Males were then necropsied and evaluated for sperm parameters and reproductive histopathology. Females continued to be dosed through lactation. Some dams and pups were necropsied at weaning (day 21); others were treated until the pups were 28 or 45 d old. No treatment-related abnormalities were noted for any of the described groups.

## D. Mutagenicity

No references to the evaluation of the mutagenicity of chlorine gas have been found. A related issue is the generation of mutagens in the treatment of food and food processing. Chlorine process water is used in large quantities in the food-processing industry to wash meat carcasses, seafood, vegetables, and fruits. Because these food products contain large amounts of amino acids, proteins, and lipids, mutagenic reaction products may be produced as a result of reacting with chlorine at high concentrations. Chlorine is used in commercial poultry processing. Eviscerated carcasses are immediately cooled in chlorinated chiller water to improve meat quality. Chlorine is added to the chiller water to obtain concentrations from a few up to 20 ppm. Some operations employ "super-chlorination," enough chlorine being added to maintain 5 ppm in discharge overflow (Masri 1986).

Masri (1986) evaluated the effect of chlorinated poultry chiller water, obtained from two separate processing plants, on the production of mutagens and the kinetics of the disappearance of added chlorine. At low chlorine levels, chlorine disappeared quickly but there was little mutagenic activity associated with the chiller water (100 ppm and less) in the Ames *Salmonella* mutagenicity assay (*Salmonella typhimurium* strain TA100 without metabolic activation). At concentrations greater than 250 ppm, mutagenic activity rose sharply and a dose–response relationship was demonstrated. Chiller water containing levels of chlorine typical in these applications (3–20 ppm) was devoid of mutagenic activity. When excess chlorine was added to the chiller water, a mutagenic response was evident, demonstrating the detection capability of the assay. The authors observed that chlorine induced no mutagenic activity in chiller water when used at levels typical for this application, and concluded that there was no existing health hazard from current plant practices.

Owusu-Yaw et al. (1991) demonstrated that nonvolatile reaction products generated from the reaction of 70 mM aqueous chlorine with 10 mM L-tryptophan were direct acting mutagens to *Salmonella typhimurium* TA100 and TA98. In addition, these fractions were shown to be capable of significantly increasing the frequency of sister chromatid exchanges in Chinese hamster ovary cells in absence of S9 mix.

## IV. Environmental Effects of Chlorine

This section describes the potential effects of chlorine on the environment, including accidental releases of chlorine gas. Chlorine may affect the environment through the atmosphere as HCl gas and aerosol, or may enter aquatic systems through the use of chlorine gas for disinfection and defouling of water and sewage effluents or as chloride ion (Cl⁻) in precipitation. The ultimate chemical fate of chlorine in wastewater is as chloride (Jolley and Carpenter 1984). Chloride formed from chlorination processes (i.e., discharged chlorinated condenser cooling waters, chlorinated sewage effluent) is relatively insignificant compared to background levels (Stevens et al. 1985). The oxidation of natural and anthropogenic organics is the most important process in the decay of free chlorine in the environment. Chlorination of waters finally leads to chloride, oxidized organics ($CO_2$), chloro-organics, oxygen, and nitrogen (Dotson et al. 1986). Chlorine is not known to bioaccumulate in the environment or the food chain; however, lipophilic oxidation products can potentially accumulate in animal tissues. Current evidence for environmental accumulation of these compounds is inconclusive and not well understood.

## A. Atmospheric Exposure

Little information is available that conclusively describes environmental exposure to ambient atmospheric chlorine. The effects of chlorine levels have been studied in the stratosphere, but studies on the effects of chlorine on vegetation and animals have primarily been conducted in the laboratory. A few reports do exist on the effects of catastrophic chlorine releases on vegetation and livestock.

*Effects on Vegetation.* **Accidental and Air Pollution Exposures.** The exposure of plants to elevated levels of chlorine can result in injury. The effects of chlorine on plants has been recorded as far back as 120 yr ago in Europe, near brickworks, clay product plants, and chemical plants (NAS 1976). In recent years, chlorine-induced damage to vegetation has been linked to accidental chlorine gas leakage, such as from sewage treatment plants, train derailments, and industrial accidents, and as a result of air pollution. In 1969, Brennan et al. evaluated the damage to plant species as a result of an accidental chlorine release at a sewage treatment plant. The chlorine gas plume produced a swath of plant damage 160 m wide and approximately 485 m long. Extensive damage was seen in many species of grasses, weeds, annual flowers, fruit trees, shade trees, and evergreens.

The symptoms of phytotoxicity were evaluated 5 and 14 d postexposure. Initially the foliage of the broadleaf plants showed varied symptoms: bleaching, intercostal necrotic spots, marginal burning, and, in some, complete necrosis of the leaves. Conifers evidenced necrosis of the tips of the current season's needles, while *Arbor vitae* seemed resistant to the chlorine induced phytotoxicity. At 14 d postexposure, new growth was evident in the damaged plants. Phytotoxicity can also indicate whether a chlorine release has recently occurred. A chemical company in England was fined because it did not issue a chlorine leak warning for 7 d, until a local farmer complained about damage to wheat crops in his field next to the chemical plant (Anonymous 1996).

Vegetation acts as an important sink for chlorine air pollution. Plant exposures to elevated levels of chlorine can cause plant injury; however, chlorine tends to be converted to other less toxic forms rather rapidly in plants and may not result in the direct accumulation of toxic pollutant residuals important in the biological food chain (Bennet and Hill 1975). Field studies evaluating the effect of chlorine in air pollution near chemical manufacturing plants have been conducted by Harper and Jones (1982) and Vijayan and Bedi (1989). Harper and Jones (1982) evaluated the effects of chlorine air pollution on 50 plant species growing in the vicinity of an anhydrous aluminum chloride plant. American elm, bur oak, eastern white pine, basswood, red ash, and several bean species were the most sensitive and displayed foliar injury at concentrations of chloride ion in the whole leaves in excess of 0.2% of dry weight. The tolerant species displayed no foliar toxicity in spite of foliar concentrations of chloride ion as high as 5% (Harper and Jones 1982).

In the village of Ranoli in western India, chlorine is a major air pollutant that is attributed to a chemical manufacturing plant. Vijayan and Bedi (1989) evaluated the effect of chlorine pollution on three economically important tropical evergreen fruit trees—mango, rayan, and jamon. They compared an impacted area within a 2.5-km radius around the chemical plant and a nonimpacted, control area located 36 km away from the pollution source. The parameters evaluated included mean leaf area, fruit yield, and biochemical parameters (chlorophyll *a* and *b* pigments, protein, carbohydrate, and chloride content of foliar tissues). One hundred trees of each type were evaluated in the pollution and control zones. Those trees evaluated in the areas closest to the chemical plant demonstrated a decrease in mean leaf area, increase in percent damaged leaf area, decrease in fruit yield, chlorophyll pigments, protein, and carbohydrate content, and an increase in chloride accumulation in foliar tissue, especially in the mango and jamon. The order of sensitivity of these fruit trees to the chlorine pollution was mango, jamon, and rayan.

**Experimental Exposures.**  Although the chlorine concentration for most field observations have not been quantified, numerous experimental studies have been conducted using known concentrations of chlorine gas. Brennan et al. (1965, 1966) have conducted numerous experiments on the effect of ambient concentrations of chlorine gas on plant life. Chlorine is usually present ambient air when used directly or as a by-product of chemical or manufacturing processes. Chlorine fumigations of many plant species were conducted at concentrations ranging from 0.1 to 1.5 ppm for 4-hr exposure periods (Brennan et al. 1965). The plant species evaluated included alfalfa, begonia, tomato, tobacco, radish, cucumber, and pepper. Some plants were harvested immediately for chloride concentration while others were held in observation for 1 wk. There was considerable variation in sensitivity to chlorine and in symptom expression following exposure to a toxic dose. Necrosis and bleaching of leaves was the most common effect. In addition, it appeared that the older leaves were more sensitive than younger leaves and that the upper leaf surface was more sensitive to the gas exposure. In terms of species sensitivity, alfalfa and radish were the most sensitive plants tested; the threshold exposure causing damage was 0.1 ppm for a 2-hr exposure. Begonia and pepper plants were among the least sensitive to chlorine toxicity. It did not appear that the chloride content of the tissues was related to the degree of chlorine exposure or to the amount of damage produced by the gas in these short-term exposures.

Brennan et al. (1965) reported that exposure of the plants to an atmosphere of chlorine did not alter the pattern of chloride distribution within the plant, and there was no appreciable increase in chloride content of any of the fractions after fumigation. An evaluation of the effects of moisture, water stress, and prolonged darkness on chlorine toxicity in these plants was also conducted. Moisture had no effect on the extent of chlorine damage, whereas prolonged darkness appeared to reduce the extent of injury if the shade period followed fumigation. If the darkness period preceded fumigation, there was no reduction in chlorine-induced injury. Water stress, as well as closing of the stomata by artificial means, had a tendency of reducing sensitivity to chlorine gas. Brennan et al. (1966) also evaluated the response of pine trees to chlorine in the atmosphere. Three species of pine (shortleaf, slash, and loblolly) were fumigated with chlorine and evidenced visible injury after a 3-hr exposure to 1 ppm chlorine. Although damage increased as the concentration increased, there was no defoliation and the chlorine exposure did not kill new shoots. The chloride content of the pine needles generally increased after chlorine gas fumigation; however, the increment was not proportional to the degree of fumigation nor to the extent of damage.

Zimmerman (1955) fumigated 19 species of plants with chlorine at concentrations ranging from 0.46 to 4.67 ppm and found that 16 of the 19 were susceptible to injury with the concentration used. Those species most susceptible to chlorine included peach, coleus, cosmos, buckwheat, and hybrid tea. Chinese holly, eggplant, and tobacco had no visible injury from any of the treatments. The most characteristic response of all the injured species was spotting of the leaves, which turned straw colored or brown depending on the species involved.

Benedict and Breen (1955) investigated the use of weeds as a means of evaluating vegetation damage caused by air pollution. Ten weed species were fumigated with a low (0.5 ppm) or a high (2.5 ppm) concentration of chlorine gas for a period of 4 hr. The following plants were fumigated: annual bluegrass, cheeseweed, chickweed, dandelion, Kentucky bluegrass, lamb's-quarter, mustard, nettle-leaf goosefoot, pigweed, and sunflower. Damage appeared in the broadleaf plants as necrotic areas produced between the veins, usually developing close to the margin and spreading toward the midrib with the higher fumigation concentration. The markings in the grasses first appeared as marginal streaks progressing toward the main vein. Mustard, chickweed, and sunflower were the most sensitive and Kentucky bluegrass, lamb's-quarter, and pigweed were the least sensitive. Plants grown under low moisture conditions were less injured than those grown under high moisture conditions.

**Effect of Acid Precipitation.**   Sulfate, from $SO_2$ emissions, and nitrate, from $NO_x$ emissions, are the major contributors to atmospheric acidity (NAPAP 1987) and as components of acid deposition have been implicated in playing a role in worldwide forest decline (Hinrichsen 1987). Chlorine and HCl play minor roles in worldwide acid deposition as compared to sulfate and nitrate emissions. The major anthropogenic source of HCl to the atmosphere is HCl emitted from combustion, including coal and the incineration of plastics. When chlorine and hydrogen chloride mix in the atmosphere with oxygen and water vapor, dilute solutions of strong mineral acids are formed that fall to earth as acid rain, snow, and fog or as acidified dry particles. Acid precipitation in the pH range of 2.3–5.0 can leach potassium, calcium, and magnesium from the leaves and needles of sugar maples, yellow birch, and white spruce. In North America, acid precipitation is the fifth leading cause of low-elevation forest decline whereas in Western and Eastern Europe it is ranked second and fourth, respectively (Hinrichsen 1987).

**Effects on Animals.**   There are no data regarding the health effects of environmental exposures to historic or current atmospheric chlorine concentrations in animals. Joyner and Durel (1962) described the effects of a catastrophic exposure to domesticated animals following a spill of 6000 gal of liquid chlorine in La Barre, LA. The chlorine gas cloud that was produced covered approximately 6 square miles, resulting in the evacuation of approximately 1000 residents. Within 36 hr, hundreds of domestic animals died. In another incident, 15 t of chlorine escaped from a chemical works plant and drifted downwind, covering 25 square miles (Clarke and Clarke 1981). Cattle were most severely affected, showing dyspnea, lacrimation, profuse nasal discharge, and depression, with several animals dying within 1–2 d. Pigs showed salivation, lacrimation, coughing, vomiting, and anorexia. Horses suffered from frequent urination and produced cracking sounds in their lungs; several were dyspneic 6 mon later. The chlorine concentrations in both these incidents were not reported.

**Effects on Property.**   Materials exposed to the environment undergo physical and chemical changes that may affect either their functionality or aesthetic value. Materials are affected by many environmental constituents, including chemicals, radiation, temperature stresses, and physical stresses. For most materials, environmental pollutants are not a primary cause of degradation, but instead may influence the rate at which degradation occurs; e.g., rusting of iron occurs in the presence of oxygen and water, but may be accelerated if acidic species are also present on the surface.

Chlorine, in its gaseous, hydrochloric acid form or its aqueous, chloride form, has been documented as having a role in the corrosion of building materials including metals and mineral products. The action of corrosion depends on environmental factors such as moisture, salt, particulates, temperature, $CO_2$ levels, and microorganisms (NAPAP 1987; Ogundele 1989). Commonly used metals usually form oxides in the earth's atmosphere, thus producing corrosion. Some materials, such as paint, provide a protective coating. Accelerated corrosion takes place in a humid atmosphere. A film of water will absorb on a metal's surface to a degree that depends on the relative humidity and the affinity of the surface for water molecules. When the surface is completely clean and dry, the circuit is open, no current can flow, and no electrochemical reaction takes place. When water is adsorbed on the surface of the metal, it acts as an electrolyte, allowing current to flow, and corrosion takes place.

Chlorine accelerates corrosion by dissolving in the absorbed water film to supply electrolyte and can in fact cause adsorption of additional water by lowering the vapor pressure of the film. In addition, through changes in the acidity of the water film, the ions produced in the water film can increase the solubility of corrosion products, tending to expose fresh surfaces to attack (NAS 1976). However, no field studies to date have estimated the actual damage to existing structures from acid deposition, and none has estimated the contribution of chloride ions. Laboratory experiments using high concentrations of chlorine have been conducted to evaluate the effects on various materials.

*Metals.*   Chlorine, in low moisture situations, does not normally corrode steel. However, in the presence of moisture, iron and steel are easily corroded in the presence of hydrogen chloride (HCL, gas) and chloride ions (liquids) (Graedel and McGill 1986). In addition, many stainless steel alloys have poor resistance to chloride ions on their surfaces (NACE 1984). Chloride ion accelerates stress corrosion cracking of both high-strength steels and stainless steels. Chlorides also lead to pitting corrosion on iron, carbon steels, and stainless steels (NACE 1984). Coatings such as properly selected and maintained paint systems are effective in preventing or dramatically slowing the impact of chlorides on iron and steel, e.g., paint systems on steel bridge superstructures. Chlorine also affects aluminum and zinc, again primarily causing pitting and stress corrosion cracking. On the other hand, chlorine has little impact on silver, copper, bronze, brass, tin, nickel, or lead (Graedel and McGill 1986).

*Other materials.*   Stone and wood are generally resistant to chlorine attack. In designing environments for sensitive electronics (e.g., computers; control and recording equipment), chlorine is generally avoided, perhaps because of its corrosion impact on ferrous metals. No data were found for other materials, probably indicating that chlorine has little or no impact on other materials.

## B. Waterborne Exposure

The addition of chlorine as hypochlorite or as gaseous chlorine to cooling towers and discharged effluents is carried out several times daily in the routine operation of power-producing plants and sewage processing plants (Brooks and Baker 1972). The resultant intermittent discharge of chlorinated waters and effluents into receiving waters may be harmful to aquatic organisms. Discharge levels of approximately 2 mg/L TRC have been implicated in fish kills on the James River in Virginia (Roberts et al. 1975). The toxicity of TRC is apparently a function of FRC in the discharged effluents. Environmental variables such as pH, salinity, and temperature also play a role in the toxicity of chlorine to aquatic organisms (Turner and Chu 1981). Mattice and Zittel (1976) attempted to develop a general relationship between TRC levels and toxicity and calculated a chronic toxicity threshold of 0.0015 mg/L for freshwater organisms and 0.02 mg/L for marine organisms.

*Effects on Aquatic Organisms.*   Laboratory as well as field investigations have been conducted to evaluate the effect of chlorinated discharges into fresh, estuarine, and marine environments. Most studies measure the effect of TRC and may not identify the various chlorinated species present in the study. Laboratory experiments typically use sodium chlorite or sodium hypochlorite as the source for aqueous chlorine. Endpoints typically observed for such experiments include lethality (usually the $LC_{50}$) and avoidance concentrations. Avoidance reactions are behavioral in nature; when given the option, using a two-sided aquarium, one side chlorinated, the other not, some fish actively avoid chlorinated water.

In addition, the effect of temperature on the toxicity of chlorinated waters has also been investigated.

*Field Experiments.*   Field experiments have been conducted to assess the effect of chlorine or chlorine reaction products (usually TRC) in the environment. In a series of field studies, Tsai (Tsai 1968, 1970, 1973) evaluated the effect of chlorinated sewage effluents on fish populations and migrations and on water below sewage outfalls. This series of studies showed a drastic decrease in number of species and fish abundance, a decrease in water quality, and an impact on odor and the physical and biological appearance of stream bottoms as a result of chlorinated sewage. In 1984, Osbourn correlated the level of aquatic organism diversity with TRC. As chlorination increased, the structure of the downstream macroinvertebrate communities significantly decreased in the Sheep River, Alberta, as measured by the Shannon–Weaver diversity index. Table 6 summarizes selected toxicity studies and the associated $LC_{50}$ or avoidance concentrations obtained in controlled field–laboratory studies. In the avoidance tests, the fish appear to be more sensitive to the free chlorine than total residual chlorine. This observation corresponds to similar findings that have been reported from laboratory experiments.

*Laboratory Experiments.*   Laboratory experiments have shown that chlorine is toxic to may aquatic organisms including microbial, macrophytic, invertebrate, and vertebrate species. Because of this property, chlorine is used as a biocide in numerous applications.

*Microbial Communities.*   Carpenter et al. (1972) evaluated the effect of chlorine added as a gas to cooling tower water on freshwater phytoplankton activity at the outfall. The highest concentrations tested were 1.2 ppm in the cooling tower water and 0.4 ppm at the outfall. These test conditions resulted in an 83% decrease in phytoplankton activity. The lowest concentration tested in the cooling tower water was 0.1 ppm, resulting in an unmeasurable concentration at the outfall. A 79% decrease in phytoplankton activity was observed under these lower concentrations. Carpenter et al. (1972) concluded from these studies that chlorine cannot be used effectively as a biocide for fouling organisms without having an adverse effect on entrained freshwater phytoplankton. However, it would appear that the toxicity would be significantly reduced by removal of chlorine before discharge. Chlorinated seawater, from open coast or estuarine environments, does not appear to be as toxic to marine phytoplankton. Coughlan and Davis (1981) evaluated the effect of chlorination on zooplankton and phytoplankton at several power stations, three stations that discharged chlorinated effluent into estuaries and two which were located on the open coast. Coughlan and Davis compared inlet samples (no chlorine) to outlet samples (after chlorination). In the absence of chlorine or at levels below 0.2 mg/L, there was negligible mortality in the adult calanoid (zooplankton) test population when compared with the inlet samples. Mortality increased linearly between 0.2 and 1.5

Table 6. Summary of selected field–laboratory studies.

| Common name | Chlorine type | Exposure time (hr) | Avoidance (A)[a] or LC$_{50}$ (L) concentration (mg/L) | Reference |
|---|---|---|---|---|
| Brown trout (caged) | Chlorinated discharge | 48 hr × 2 Cl events | 0.14–0.17 (L) | Basch and Truchan 1976 |
| | | 48 hr × 4 Cl events | 0.18–0.19 (L) | |
| | | 96 hr × 3 Cl events | 0.02–0.05 (L) | |
| | | 96 hr × 6 Cl events | 0.17–0.18 (L) | |
| Rainbow trout | Chlorinated discharge: TRC | 2 hr | 0.15, 24% (A) | Schumacher and Ney 1980 |
| | | | 0.25, 45% (A) | |
| | | | 0.50, 95% (A) | |
| Spotfish shiner | Chlorinated discharge: FRC | NA | 0.06, 50% (A) | Cherry et al. 1977 |
| | | | ≥0.27, 100% (A) | |
| Bluntnose minnow | Chlorinated discharge: FRC | NA | 0.06, 50% (A) | Cherry et al. 1977 |
| | | | ≥0.27, 100% (A) | |

[a] Avoidance response; percent of fish retreating from associated TRC/FRC levels. TRC, total residual chlorine; FRC, free residual chlorine.

mg/L TRC. At one site, a 2.2 mg/L concentration produced 25% more mortality than was predicted from the linear model. At another site, mortality was substantially higher than the extrapolated values. Toxicity to phytoplankton was estimated from the reduction of photosynthetic activity between the inlet and outlet samples, under standard conditions of incubation.

The dose–response relationship in phytoplankton was not linear but could be described by an exponential curve or by a power curve. The effect of chlorine was also characterized by a 3-hr $EC_{50}$, which is the concentration of residual chlorine necessary to achieve a 50% reduction in productivity after a 3-hr incubation. The three estuarine sites had comparable $EC_{50}$ values (0.13, 0.12, and 0.11 mg/L), while one open coast plant had an $EC_{50}$ value of 0.29 mg/L, twice as high. At the remaining open coast plant, either model (linear or exponential) used predicted a negative chlorine value for the $EC_{50}$. On further examination, it was determined that the concentration of chlorine at the outlet was substantially higher than measured in the samples.

Pratt et al. (1988) conducted laboratory tests that proved to be more sensitive than field evaluations. Under constant exposure, algal biomass was adversely affected at concentrations as low as 2 μg/L. Alkaline phosphatase activity was inhibited at 6 μg/L whereas oxygen production and other biomass measures were affected at 0.025 mg/L and 0.025–0.3 mg/L, respectively. Observations from their field enclosure experiments included decreased protozoan numbers at pulse exposures of 0.079 mg/L, decreased zooplankton density at 0.024 mg/L, and both algal and total biomass reductions at 0.261 mg/L.

*Invertebrates.* Aquatic invertebrates may be the most sensitive aquatic species to chlorine and chlorinated reaction products in the water. Assessment of the effects of chlorine and chlorination products in the water depends on the chlorine form (free or residual) as well as the individual organism and its stage of development. Molluscan larvae are among the most sensitive species and developmental stage, with $LC_{50}$ concentrations of only 0.005 mg/L (Roberts et al. 1975). Chlorine residuals have also been shown to adversely effect reproductive rates and egg production of rotifers (Capuzzo 1979) and to decrease fertilization of sea urchins, echiuroids, and annelids at concentrations as low as 0.05 mg/L available chlorine (Muchmore and Eppel 1973). Table 7 summarizes the selected toxicity data for some of the invertebrates. The magnitude of toxicity appears to depend on life cycle stage (early stages are more sensitive than adult), temperature (chlorine and high temperature are more toxic than chlorine and lower temperatures), length of exposure, and species.

*Fish.* Fish exhibit toxic responses to chlorine and chlorine reaction products. Fish also exhibit an avoidance response to chlorine and chlorine reaction products. The response of fish has been characterized as a threshold effect (Capuzzo et al. 1977), with toxicity increasing abruptly over a narrow concentration range.

Studies have been conducted to evaluate the behavioral alterations and physiological effects of chlorine-exposed fish. Larson et al. (1978) described the be-

Table 7. Summary of selected chlorine toxicity data on invertebrates.

| Species or common name | Life stage | Chlorine type | Exposure (time hr/temp °C) | Effect:[a] Lethality (L) or reproduction (R) (mg/L) | Reference |
|---|---|---|---|---|---|
| *Cyclops bicuspidatus thomasi* | Adult | TRC | 0.5/10° | 14.68 (L-50%) | Seegert et al. 1977 |
| | Adult | TRC | 0.5/15° | 15.61 (L-50%) | Seegert et al. 1977 |
| | Adult | TRC | 0.5/20° | 5.76 (L-50%) | Seegert et al. 1977 |
| *Limnocalanus marcrurus* | Adult | TRC | 0.5/10° | 1.54 (L-50%) | Seegert et al. 1977 |
| Brine shrimp | Nauplii | TRC | —/— | 0.047 (L-70%) | Puente et al. 1992 |
| | Adult | TRC | —/— | >0.095 (L-50%) | Puente et al. 1992 |
| Glass shrimp | Adult | TRC | 96/— | 0.22 (L-50%) | Bellanca and Bailey 1977 |
| *Daphnia magna* | Adult | TRC | 0.5/— | 0.097 (L-50%) | Mattice et al. 1981 |
| | Adult | TRC | 1/— | 0.063 (L-50%) | Mattice et al. 1981 |
| American lobster | Larva | FRC | 0.5-1/25° | 16.30 (L-50%) | Capuzzo et al. 1976 |
| | Larva | FRC | 0.5-1/30° | 2.50 (L-50%) | Capuzzo et al. 1976 |
| Mud crab | Adult | CPO[b] | 96/25° | 1.06 (L-50%) | Key and Scott 1986 |
| Copepod species | Larva | TRC | 48/— | 0.005 (L-50%) | Bellanca and Bailey 1977 |
| Oyster | Larva | TRC | 48/— | 0.005 (L-50%) | Bellanca and Bailey 1977 |
| Clam | Larva | TRC | 48/— | 0.005 (L-50%) | Bellanca and Bailey 1977 |
| Blind cave crayfish | Adult | FRC | 24/— | 2.25[c] (L-50%) | Matthews et al. 1977 |
| | Adult | FRC | 24/— | 3.00[c] (L-50%) | Matthews et al. 1977 |
| *Daphnia magna* | Adult | TRC | 72/— | 0.5 (L-100%) | Ellis 1937 |
| | Adult | TRC | 336/— | 0.002 (R-decreased) | Arthur et al. 1975 |
| *Gammarus pseudolimnaeus* | Adult | TRC | 48/15° | 0.023 (L-50%) | Gregg 1974 |
| | | | 96/— | 0.22 (L-50%) | Arthur and Eaton 1971 |
| | | | 2520/— | 0.0034 (R-100%) | Arthur et al. 1975 |
| *Nais* sp. | Adult | TRC | 0.5/— | 1.0 (L-95%—100%) | Collins 1958; Learner and Edwards 1963 |
| *Trilobus gracilus* | Adult | TRC | 2.5/— | 20.0 (L-100%) | Collins 1958 |
| | Juv. | TRC | 1.5/— | 3.0 (L-100%) | Collins 1958 |

—/—, not reported; [a]L or R, x %; percent of lethality or decrease in reproduction.
[b]CPO, Chlorine-produced oxidants, equivalent to TRC; [c]Crayfish were evaluated before and after a 3-d acclimation period to FRC (0.01–1.20 mg/l).

havioral alterations of fish after exposure to concentrations near median lethal concentrations of chlorine and recorded the phases of toxicity with increasing concentration. The observations included changes in rates of swimming, opercular movement, and breathing, disorientation, and, at acutely toxic concentrations, "resting" on the bottom of aquaria, followed by floating belly up. Those fish not dying from the chlorine exposure usually recovered within 24 hr after exposure. Booth et al. (1981) exposed fish to 0.11 mg/L and 0.22 mg/L for 2 hr, followed by 6 hr of dechlorinated water. The fish experienced increased coughing and subcutaneous hemorrhaging, the severity of which was dose related. The ventilation rates decreased with an increase in chlorine concentration. The 0.22 mg/L exposed fish all died within 2 hr following the experiment. The moribund fish moved to the top of the tank, gasping, indicating hypoxia. Bass and Heath (1975) evaluated the physiological effects of chlorination on rainbow trout. Rainbow trout were exposed intermittently three times per day to chlorinated water to simulate conditions below steam electricity generating plants. During each chlorination $pO_2$, pH, and heart rate decreased while breathing and coughing rates increased. Partial recovery of all parameters occurred after each pulse of chlorine; however, the amount of recovery decreased with each successive chlorination until death occurred. Pathological and histological evaluations revealed an increase in mucous production and damage to the respiratory epithelium. It was concluded that the mode of action was gill damage resulting in death by asphyxiation.

The effect of chlorine on the early life stages of fish has been investigated. In general, it appears that the eggs are more tolerant to TRC concentrations than freshly hatched larvae, and tolerance increases as development proceeds to metamorphosis (Alderson 1974; Middaugh et al. 1977b; Morgan and Prince 1977; Burton et al. 1979; Yosha and Cohen 1979). In addition, Morgan and Prince (1977) found that exposing blackfish eggs to increasing concentrations of TRC resulted in abnormal larvae. Blackfish eggs exposed to TRC levels below 0.26 ppm evidenced obvious anomalies of the vertebral column in 1.6% of the hatchlings. At 0.31–0.38 ppm, approximately 15% of the larvae were abnormal.

Numerous investigations have been conducted evaluating sublethal and lethal effects of chlorine on fish under various exposure conditions and life stages. Table 8 summarizes selected toxicity studies and the associated $LC_{50}$ or avoidance concentrations obtained in these laboratory experiments. The effects vary depending on chlorine form, temperature of the environment, and the exposure regimen. Toxicity is both species- and developmental stage dependent, with larval stages most sensitive. A number of investigators have reported that free chlorine (HOCl or OCl⁻) is more toxic than total residual chlorine and that HOCl is the most toxic species, having a toxicity approximately two- to sixfold greater than that of OCl⁻.

*Effects on Aquatic Vegetation.* Although the effect of aqueous chlorine on aquatic animals has been extensively studied, little information is available re-

Table 8. Summary of selected chlorine toxicity data on fish.

| Species (common name) | Life stage[a] | Chlorine type | Exposure[b] (time hr/ temp °C) | Avoidance (A) or LC$_{50}$ (L): concentration (mg/L) | Reference |
|---|---|---|---|---|---|
| Bluegill | A | HOCl | —/— | A-0.1–0.4 | Cherry et al. 1982 |
| | A | TRC | 96/— | L-0.4–0.45 | Bass and Heath 1977 |
| | A | TRC | 96/— | L-2.13 | Wilde et al. 1983a |
| | A | TRC | 96 I/— | L-0.88 | Wilde et al. 1983a |
| | J | TRC | 96 I/— | L-0.44 | Wilde et al. 1983b |
| Freshwater coho salmon | A | HOCl | —/— | A-0.05–0.01 | Cherry et al. 1982 |
| | A | TRC | 0.5/20° | L-0.29 | Seegert and Brooks 1978 |
| | A | TRC | 0.5/10° | L-0.56 | Seegert and Brooks 1978 |
| | A | TRC | 0.5/10° | L-1.25 | Seegert et al. 1977 |
| Rainbow trout | A | HOCl | —/— | A-0.05–0.1 | Cherry et al. 1982 |
| | Y | TRC | 0.5/10° | L-2.0 | Seegert et al. 1977 |
| | A | TRC | 0.5/10° | L-0.99 | Brooks and Seegert 1977 |
| | A | TRC | 0.5/15° | L-0.94 | Brooks and Seegert 1977 |
| | A | TRC | 0.5/20° | L-0.43–0.6 | Brooks and Seegert 1977 |
| | A | TRC | 3 × 5″/10° | L-2.87 | Brooks and Seegert 1977 |
| | A | TRC | 3 × 5″/20° | L-1.65 | Brooks and Seegert 1977 |
| | A | HOCl | 0.25/— | L-0.64 | Brooks and Bartos 1984 |
| | A | HOCl | 0.5/— | L-0.46 | Brooks and Bartos 1984 |
| | A | HOCl | 2/— | L-0.20 | Brooks and Bartos 1984 |
| | A | HOCl | 4 × 0.5/— | L-0.25 | Brooks and Bartos 1984 |
| Largemouth bass | A | HOCl | —/— | A-0.1–0.4 | Cherry et al. 1982 |
| Mosquito fish | A | HOCl | —/— | A-0.1–0.4 | Cherry et al. 1982 |
| Fathead minnow | A | TRC | 96/— | L-0.07–0.19 | Zillich 1972 |
| | J | TRC | 96/— | L-0.39 | Wilde et al. 1983a |
| | J | TRC | 96 I/— | L-0.18 | Wilde et al. 1983a |
| | A | TRC | 96/— | L-1.37 | Wilde et al. 1983a |
| | A | TRC | 96 I/— | L-0.58 | Wilde et al. 1983a |
| | J | TRC | I | L-0.08 | Wilde et al. 1983b |
| | A | TRC | I | L-0.35 | Wilde et al. 1983b |
| Alewife | A | TRC | 0.5/10° | L-2.25 | Seegert et al. 1977 |
| | A | TRC | 0.5/15° | L-2.27 | Seegert and Brooks 1978 |
| | A | TRC | 0.5/30° | L-0.30 | Seegert and Brooks 1978 |
| Goldfish | A | TRC | 24/— | L-0.27 | Dickson et al. 1977 |
| | A | TRC | 12/— | L-0.4 | Dickson et al. 1977 |

Table 8. (Continued).

| Species (common name) | Life stage[a] | Chlorine type | Exposure[b] (time hr/ temp °C) | Avoidance (A) or LC$_{50}$ (L): concentration (mg/L) | Reference |
|---|---|---|---|---|---|
| Brook trout | A | TRC | 96/10° | L-0.15–0.18 | Thatcher et al. 1976 |
| | A | TRC | 96/15° | L-0.13–0.16 | Thatcher et al. 1976 |
| | A | TRC | 96/20° | L-0.10–0.12 | Thatcher et al. 1976 |
| Spottail shiner | A | TRC | 0.5/10° | L-2.41 | Seegert and Brooks 1978 |
| | A | TRC | 0.5/20° | L-0.53 | Seegert and Brooks 1978 |
| | A | TRC | 0.5/10° | L-3.2 | Seegert et al. 1977 |
| Rainbow smelt | A | TRC | 0.5/10° | L-1.27 | Seegert and Brooks 1978 |
| Emerald shiner | J | TRC | 96/10° | L-1.32 | Fandrei and Collins 1979 |
| | J | TRC | 96/25° | L-0.33 | Fandrei and Collins 1979 |
| | Y | TRC | 96/10° | L-0.71 | Fandrei and Collins 1979 |
| | Y | TRC | 96/25° | L-0.23 | Fandrei and Collins 1979 |
| | A | TRC | 96/10° | L-0.87 | Fandrei and Collins 1979 |
| | A | TRC | 96/25° | L-0.28 | Fandrei and Collins 1979 |
| | A | HOCl | 0.25/— | L-0.26 | Brooks and Bartos 1984 |
| | A | HOCl | 0.5/— | L-0.18 | Brooks and Bartos 1984 |
| | A | HOCl | 2/— | L-0.10 | Brooks and Bartos 1984 |
| | A | HOCl | 4 × 0.5/— | L-0.10 | Brooks and Bartos 1984 |
| | A | OCl$^-$ | 0.25/— | L-0.57 | Brooks and Bartos 1984 |
| | A | OCl$^-$ | 0.5/— | L-0.44 | Brooks and Bartos 1984 |
| | A | OCl$^-$ | 2/— | L-0.12 | Brooks and Bartos 1984 |
| | A | OCl$^-$ | 4 × 0.5/— | L-0.16 | Brooks and Bartos 1984 |
| Yellow perch | A | TRC | 0.5/10° | L-7.7 | Seegert et al. 1977 |
| | A | TRC | 0.5/15° | L-4.0 | Seegert et al. 1977 |
| | A | TRC | 0.5/20° | L-1.1 | Seegert et al. 1977 |
| | A | TRC | 0.5/25° | L-1.0 | Seegert et al. 1977 |
| | A | TRC | 0.5/10° | L-8 | Brooks and Seegert 1977 |
| | A | TRC | 0.5/30° | L-0.7 | Brooks and Seegert 1977 |
| | A | TRC | 3 × 5″/10° | L-22.6 | Brooks and Seegert 1977 |
| | A | TRC | 3 × 5″/20° | L-9 | Brooks and Seegert 1977 |
| Channel catfish | A | HOCl | 0.25/— | L-0.26 | Brooks and Bartos 1984 |
| | A | HOCl | 0.5/— | L-0.21 | Brooks and Bartos 1984 |
| | A | HOCl | 2/— | L-0.18 | Brooks and Bartos 1984 |
| | A | HOCl | 4 × 0.5/— | L-0.15 | Brooks and Bartos 1984 |
| | A | OCl$^-$ | 0.25/— | L-1.27 | Brooks and Bartos 1984 |
| | A | OCl$^-$ | 0.5/— | L-1.12 | Brooks and Bartos 1984 |
| | A | OCl$^-$ | 2/— | L-0.31 | Brooks and Bartos 1984 |
| | A | OCl$^-$ | 4 × 0.5/— | L-0.23 | Brooks and Bartos 1984 |

Table 8. (Continued).

| Species (common name) | Life stage[a] | Chlorine type | Exposure[b] (time hr/ temp °C) | Avoidance (A) or $LC_{50}$ (L): concentration (mg/L) | Reference |
|---|---|---|---|---|---|
| Saltwater | A | TRC | 1/13° | L-0.208 | Stober et al. 1980 |
| coho salmon | A | TRC | 1/20° | L-0.130 | Stober et al. 1980 |
|  | A | TRC | —/— | A-0.002 | Stober et al. 1980 |
| Shiner perch | A | TRC | 60/13° | L-0.308 | Stober et al. 1980 |
|  | A | TRC | 60/20° | L-0.230 | Stober et al. 1980 |
|  | A | TRC | —/— | L-0.175 | Stober et al. 1980 |
|  | A | TRC | —/— | L-0.19 | Dinnel et al. 1979 |
| Atlantic | J | TRC | 1.67/25° | A-0.15 | Hall et al. 1981 |
| menhaden | J | TRC | —/25° | A-0.1, 0.15 | Hall et al. 1982a |
| Winter | J | TRC | 0.5/— | L-0.55 | Capuzzo et al. 1977 |
| flounder | A | TRC | 0.16/25° | $LC_{100}$-0.3 | Hoss et al. 1975 |
| Striped mullet | J | TRC | 96/— | L-0.212 | Venkataramiah et al. 1981 |
|  | Y | TRC | 96/— | L-0.607 | Venkataramiah et al. 1981 |
|  | J | TRC | 0.12/35° | $LC_{100}$-0.5 | Hoss et al. 1975 |
| Blue rockfish | A | TRC | 0.33/18° | L-0.59 | Wiley 1981 |
| Spot | J | TRC | —/10° | L-0.12 | Middaugh et al. 1977a |
|  | J | TRC | —/15° | L-0.06 | Middaugh et al. 1977a |
| Striped bass | L | TRC | 24/— | L-0.19 | Hall et al. 1982b |
|  | L | TRC | 48/— | L-0.16 | Hall et al. 1982b |
|  | L | TRC | 96/— | L-0.14 | Hall et al. 1982b |
|  | J | TRC | 24/— | L-0.21 | Hall et al. 1982b |
|  | J | TRC | 48/— | L-0.20 | Hall et al. 1982b |
|  | J | TRC | 96/— | L-0.19 | Hall et al. 1982b |
|  | Y | TRC | 24/— | L-0.29 | Hall et al. 1982b |
|  | Y | TRC | 48/— | L-0.25 | Hall et al. 1982b |
|  | Y | TRC | 96/— | L-0.23 | Hall et al. 1982b |
| Scup | J | TRC | 0.5/— | L-0.65 | Capuzzo et al. 1977 |
| Killifish | J | TRC | 0.5/25° | L-0.65 | Capuzzo et al. 1977 |
|  | J | TRC | 0.5/30° | L-0.25 | Capuzzo et al. 1977 |
| Blacksmith | J | TRC | —/— | A-0.35 | Hose and Stoffel 1980 |

—/—, not reported; [a]Life Stages: A, adult; J, juvenile; Y, yearling; L, larva.
[b]Exposure: I, intermittent.

garding the effect of chlorine on aquatic plants. Watkins and Hammerschlag (1984) evaluated the extent of chlorine toxicity to a common submerged aquatic plant, the Eurasian water milfoil. Municipal water was aged and filtered through an activated carbon filter to remove chlorine, ammonium ion, and other materials. Chlorine was then added by bubbling commercially available chlorine gas into distilled water. Plants were acquired from a Potomac estuary, rerooted in appropriate soil, and acclimated to the aquarium used in the study. Plants were exposed to 0, 0.02, 0.05, 0.1, 0.3, 0.5, or 1.0 mg/L TRC for either a continuous 96-hr period or on an intermittent basis, $3 \times 2$ hr/d for 96 hr. The continuous exposure to chlorine concentrations as low as 0.05 mg/L was toxic to the milfoil. A significant decrease in shoot length, shoot and total plant dry weight, and chlorophyll content was observed. Greater reductions were noted at higher concentrations of chlorine. Necrotic discoloration was noted within 6 hr in the 1.0 mg/L group. Intermittent exposure of plants for $3 \times 2$ hr/d for 96 hr indicated an insensitivity to repeated short-term chlorine exposures at all concentrations except 1.0 mg/L. At 1.0 mg/L, a decrease in length, root dry weight, and chlorophyll content was observed.

## V. Regulatory Overview
### A. Occupational Exposure Limits

The occupational exposure limits for chlorine in the air of workplaces vary in different countries. Table 9 provides a listing of available occupational exposure limits/guidelines for chlorine. Bulgaria, Colombia, Jordan, Korea, New Zealand, Singapore, and Vietnam have adopted the ACGIH-TLV (RTECS 1997).

### B. Emergency Release Guidelines

The NRC (National Research Council), NIOSH (National Institute for Occupational Safety and Health), AIHA (American Industrial Hygiene Association), and most recently the USEPA have derived short-term, one-time-only exposure levels during chemical emergency situations. Table 10 summarizes these values.

In 1984, the NRC proposed emergency exposure limits (EELs) based upon concentrations that produced eye and nasal irritation (NAS 1984). NIOSH has established an IDLH (immediately dangerous to life and health) value for chlorine. An IDLH exposure condition is a condition "that poses a threat of exposure to airborne contaminants when that exposure is likely to cause death or immediate or delayed permanent adverse health effects or prevent escape from such an environment" (NIOSH 1987). The purpose of an IDLH exposure concentration is to ensure that the worker can escape from a given contaminated environment in the event of failure of the respiratory protection equipment (NIOSH 1994). The AIHA's short-term, emergency exposure levels are the Emergency Response Planning Guidelines (ERPGs). The ERPGs were established as guidelines for evacuations in catastrophic emergency situations. They range from a minor exposure (ERPG-1), which should only result in a mild transient adverse

Table 9. Summary of occupational exposure limits for chlorine.

| Country | Limit value (ppm) | Limit type | Governing agency | Reference |
|---|---|---|---|---|
| USA | 1 | PEL ceiling | OSHA | CFR 1997a |
| | 0.5 | TWA | ACGIH | ACGIH 1997 |
| | 1 | STEL | ACGIH | ACGIH 1997 |
| | 0.5 | 15-min ceiling | NIOSH | NIOSH 1994 |
| Arab Republic of Egypt | 1 | TWA | NA | RTECS 1997 |
| Australia | 1 | TWA | NA | RTECS 1997 |
| Austria | 1 | TWA | NA | RTECS 1997 |
| Belgium | 0.5 | TWA | NA | RTECS 1997 |
| | 1 | STEL | NA | RTECS 1997 |
| Denmark | 0.5 | TWA | NA | RTECS 1997 |
| Finland | 0.5 | TWA | NA | RTECS 1997 |
| | 1 | STEL | NA | RTECS 1997 |
| France | 1 | STEL | NA | RTECS 1997 |
| Germany | 0.5 | TWA | NA | RTECS 1997 |
| Hungry | 0.3 | STEL | NA | RTECS 1997 |
| India | 1 | TWA | NA | RTECS 1997 |
| | 3 | STEL | NA | RTECS 1997 |
| Japan | 1 | TWA | NA | RTECS 1997 |
| The Netherlands | 1 | TWA | NA | RTECS 1997 |
| The Philippines | 1 | TWA | NA | RTECS 1997 |
| Poland | 0.5 | TWA | NA | RTECS 1997 |
| Russia | 0.3 | STEL | NA | RTECS 1997 |
| Sweden | 0.5 | TWA | NA | RTECS 1997 |
| | 1 | STEL | NA | RTECS 1997 |
| Switzerland | 0.5 | TWA | NA | RTECS 1997 |
| | 1 | STEL | NA | RTECS 1997 |
| Thailand | 1 | TWA | NA | RTECS 1997 |
| Turkey | 1 | TWA | NA | RTECS 1997 |
| United Kingdom | 1 | TWA | NA | RTECS 1997 |
| | 3 | STEL | NA | RTECS 1997 |

PEL, permissible exposure limit; TWA, time-weighted average; STEL, short-term exposure limit; TLV, threshold limit value.

health effect (i.e., irritation), to a major exposure (ERPG-3) below which exposure of nearly all individuals for 1 hr would not result in death of the individual. Recently, the USEPA (1997) has proposed Acute Exposure Guideline Levels (AEGLs) for several hazardous substances including chlorine. The AEGLs and supplementary qualitative information on the hazardous substances were developed to assist federal, state, and local agencies and organizations in the private sector concerned with chemical emergency planning, prevention, and response.

Table 10. Summary of short-term exposure levels.

| Agency | Guideline | Exposure level | Rationale |
|---|---|---|---|
| NAS 1984 | EELs | 3 ppm-60 min<br>0.5 ppm-24 hr<br>0.1 ppm-90 d | Based upon concentrations that produced nasal and eye irritation |
| NIOSH 1994 | IDLH | 10 ppm | Based upon a concentration "that poses a threat of exposure to airborne contaminants when that exposure is likely to cause death or immediate or delayed permanent adverse health effects or prevent escape from such an environment" (NIOSH 1987) |
| AIHA 1996 | ERPG-3 | 20 ppm-1 hr | The maximum airborne concentration below which it is believed that nearly all individuals could be exposed for up to 1 hr without experiencing or developing life-threatening health effects |
|  | ERPG-2 | 3ppm-1 hr | The maximum airborne concentration below which it is believed that nearly all individuals could be exposed for up to 1 hr without experiencing or developing irreversible or other serious health effects or symptoms which could impair an individual's ability to take protective action |
|  | ERPG-1 | 1 ppm-1hr | The maximum airborne concentration below which it is believed that nearly all individuals could be exposed for up to 1 hr without experiencing other than mild, transient adverse health effects or without perceiving a clearly defined objectionable odor |
| USEPA (proposed 10/30/97 Federal Reg.) | AEGL-1 | 1.4 ppm-30 min<br>1.0 ppm-1 hr<br>0.5 ppm-4 hr<br>0.5 ppm-8 hr | Based upon human exposure experiments (Rotman et al. 1983); no sensory irritation, some transient pulmonary function changes |
|  | AEGL-2 | 2.8 ppm-30 min<br>2.0 ppm-1 hr<br>1.0 ppm-4 hr<br>0.7 ppm-8 hr | Based upon human exposure experiments (Rotman et al. 1983); transient pulmonary function changes, sensitive individual experienced an asthmatic attack |
|  | AEGL-3 | 28 ppm-30 min<br>20 ppm-1 hr<br>10 ppm-4 hr<br>7.1 ppm-8 hr | Based on animal lethality data (MacEwen and Vernot 1972; Zwart and Wouterson 1988); based upon a 1-hr concentration of 200 ppm divided by a combined uncertainty factor of 10 |

## C. Water Standards

Chlorine is designated as a hazardous substance under Section 311(b)(2)(A) of the Federal Water Pollution Control Act and further regulated by the Clean Water Act Amendments of 1977 and 1978. These regulations apply to discharges of this substance. Freshwater aquatic organisms and their uses should not be adversely affected if the 4-d average concentration of total residual chlorine does not exceed 11 µg/L more than once every 3 yr on the average and if the 1-hr average concentration does not exceed 19 µg/L more than once every 3 yr on the average (USEPA 1985). Saltwater aquatic organisms and their uses should not be adversely affected if the 4-d average concentration of chlorine produced oxidants does not exceed 7.5 µg/L more than once every 3 yr on the average and if the 1 hr average does not exceed 13 µg/L more than once every 3 yr on the average (USEPA 1985).

The Safe Drinking Water Act (SDWA) Amendments of 1996 have made sweeping changes to the existing SDWA. Under this law, in December 1998, EPA established Maximum Contaminant Levels (MCLs) for those disinfectant by-products that may have an adverse effect on human health, occur in a public water system at a frequency and concentration of significance to public health, and whose regulation offers a meaningful opportunity to reduce health risk for people served by public water systems. In addition, EPA established Maximum Residual Disinfection Levels (MRDL) for the disinfectants themselves. In December 1998, EPA established a MRDL for chlorine of 4 mg/L (CFR 1998).

## D. FIFRA Standards

Chlorine gas is exempted from the requirement of a tolerance when used pre- and postharvest in solution on all raw agricultural commodities (CFR 1997b). An exemption from a tolerance is granted when it appears that the total quantity of the pesticide chemical in or on all raw agricultural commodities for which it is useful under conditions of use currently prevailing or proposed will involve no hazard to the public health.

## E. FDA Standards

Chlorine may be added to flours in a quantity not more than sufficient for bleaching (CFR 1997c).

## VI. Discussion

The reactivity of chlorine makes it an ideal precursor for many industrial chemicals and as a drinking water and wastewater disinfectant. The majority of chlorine produced in the United States is used in the production of chlorinated organic and inorganic compounds. Chlorine is used in the manufacture of a variety of chemicals that are subsequently used in the manufacture of an extensive array of consumer products such as plastics and pharmaceuticals.

The use of chlorine gas in water treatment includes disinfection of potable

water, wastewater effluent treatment, and as a biocide for equipment maintenance in power generation plants, desalination plants, petrochemical plants, and
in the paint and metal industries (NAS 1976). The strong oxidizing nature of
chlorine, its ease of application, and low cost have led to its use as a primary
disinfectant. This disinfectant ability of free chlorine (HOCl/OCl⁻) is related to
the oxidizing nature of the free chlorine residual such that it can inactivate a
broad range of waterborne bacterial, viral, and protozoal pathogens. It is used
in the treatment of sewage effluent to control pathogenic organisms, reduce
odor, and reduce biochemical oxygen demand (BOD) (Laubusch 1962b). Chlorine is also used as a biocide to reduce biofouling in cooling water towers of
power plants, desalination plants, and the petrochemical, paint, and metal industries. Chlorination removes biota that form on conduits, piping, and the heat-
exchange surfaces of condensers (NAS 1976).

As discussed here, chlorine has many uses in its gaseous form or in solution.
The strong oxidizing nature of chlorine, which led to its use as a primary disinfectant, is also the primary mode of toxicity. The primary routes of exposure
are dependent on the form in which it is used. Chlorine gas is used in a number
of applications, and thus inhalation and skin and eye contact with the gas are
the primary routes of exposure. Chlorine forms both hypochlorous and hydrochloric acid on contact with moist mucous membranes. Hypochlorous acid decomposes into hypochloric acid and free oxygen radicals ($O_2^-$), which may disrupt the integrity and increase the permeability of the epithelium (WHO 1982)
and, by an increase in hydrogen ions, decrease the blood pH, provided that a
sufficient dose of chlorine gas is absorbed (Wood et al. 1987). Chlorine may
also react with the sulfhydryl groups of amino acids, thereby inhibiting various
enzymes (WHO 1982; McNulty et al. 1983). Resultant tissue damage is caused
by the disruption of cellular proteins (Ellenhorn and Barceloux 1988).

Currently, the immediate acute effects of chlorine gas are well characterized
in both humans and animal models. The clinical presentation of acute chlorine
gas poisoning is dependent on concentration of chlorine and duration of exposure as well as water content of tissue involved and the presence of underlying
cardiopulmonary disease (Ellenhorn and Barceloux 1988). In mild exposures,
clinical signs include irritation of the eyes, respiratory tract, and mucous membranes, rhinorrhea, cough, headache, sore throat, chest pain, dyspnea, nausea,
and pulmonary function deficits. After more severe exposures, clinical signs
include ulcerative tracheobronchitis, pulmonary edema, respiratory failure, and
death. Corneal abrasions and cutaneous burns from direct exposure can also
occur (Ellenhorn and Barceloux 1988).

Accidental exposures to large concentrations have been documented in the
literature. In addition to the clinical signs already described, investigators have
tracked exposed subjects over long-term periods. Charan et al. (1985) documented cases of persistent airflow obstruction and a progressive decline in pulmonary residual volume over 12 yr in workers accidentally exposed to chlorine
gas. The prevalence of persistent respiratory symptoms and bronchial hyperresponsiveness due to RAD has been documented as well (Bherer et al. 1994;
Donnelly and Fitzgerald 1990). Further research is needed to evaluate whether

acute inhalation of high concentrations of chlorine gas leads to permanent impairment of pulmonary function. In addition, just as a single acute exposure may lead to pulmonary deficits, chronic inhalation of low concentrations of chlorine gas may also permanently impair lung function or exacerbate preexisting pulmonary conditions; however, a clear correlation has not been made at this time.

Chlorine gas has not been associated with carcinogenicity, reproductive, mutagenic, or immunological toxicity in humans or animal models. There have been a few reports of neurotoxicity as the result of a catastrophic exposure to chlorine gas. The reports of chlorine neurotoxicity are not well established, and this phenomenon has not been evaluated in an animal model.

In animals, subchronic exposures to concentrations of 9 ppm have led to moderate inflammatory changes in the trachea and bronchiolar area and in the bronchioles and alveolar ducts. The pathology of rodents subchronically exposed to chlorine gas at concentrations to 3 ppm revealed inflammation of the nasal turbinates and the submuscosal epithelium of the trachea (Barrow et al. 1979). Chronic exposure to chlorine gas at concentrations to 3 ppm has led to mucous cell hyperplasia in the rodent nose (Wolf et al. 1995), whereas monkeys had focal decreases in mucous cells on the margins of the middle turbinates (Klonne et al. 1987). Leininger et al. (1994) compared the histopathology of the rhesus monkey study (Klonne et al. 1987) and the CIIT mouse and rat study (Wolf et al. 1995) to evaluate the differences in histopathological findings. The areas of confirmed injury correlates with airflow studies that concluded that these areas were at risk for toxic injury. These data clearly show species differences in areas susceptible to toxic injury by chlorine.

It is not known whether humans respond like monkeys or rodents. Generation of meaningful epidemiology or nasal toxicology data is needed to answer this question. It should be noted, however, that rodents are obligate nose breathers whereas humans and monkeys are not. This physiological difference may also impart a profound effect on toxicity of chlorine to the nasal epithelium. Work by Nodelman and Ultman (1999a,b) has begun to answer these questions. These authors have shown that exposures up to 3 ppm chlorine gas to normal, healthy, nonsmoking adults leads to 95% absorption in the upper respiratory tract. Less than 5% of the inhaled chlorine reached beyond the upper airways, and none reached the respiratory air spaces. Furthermore, these studies have shown that almost all the inhaled chlorine was absorbed in the nose during nasal breathing and in the mouth during oral breathing. Thus, it is logical to conclude that in humans the mouth and nasal cavity may be a target for long-term tissue damage induced by low concentrations of chlorine.

The environment is constantly being exposed to chlorine, whether by natural sources (e.g., photolysis of salt in seawater and volcanic eruptions) or through anthropogenic sources (e.g., industrial releases or directly into the water through water disinfection). Chlorine is not known to bioaccumulate in the environment or the food chain. The oxidation of natural and anthropogenic organics is the most important process in the decay of free chlorine in the environment. Chlorination of waters finally leads to chloride, oxidized organics, chloro-organics, oxygen, and nitrogen (Dotson et al. 1986). Current evidence for environmental accumulation of these compounds is inconclusive and not well understood.

Vegetation acts as an important sink for chlorine air pollution. Although plant exposures to elevated concentrations of chlorine can cause injury, plants tend to convert chlorine to other less toxic forms, therefore decreasing the direct accumulation of toxic pollutant residuals that may be detrimental to the biological food chain. Currently, there is little information regarding exposure of wildlife to ambient levels of atmospheric chlorine. These have been incidences of catastrophic exposures to domestic animals following spills. Extensive work has been conducted on the effect of waterborne chlorine exposures on aquatic animal life. Chlorine is constantly being introduced to waterways via discharged effluents in the routine operation of power-producing plants and sewage processing plants. The resultant intermittent discharge of chlorinated waters and effluents into receiving waters may be harmful to aquatic organisms. This toxicity is apparently a function of free residual chlorine in the discharged effluents; however, environmental variables such as pH, salinity, and temperature also play a role in the toxicity of chlorine to aquatic organisms (Turner and Chu 1981).

## Summary

This review provides a summary of information available on human health effects, environmental fate, and environmental toxicology of chlorine. Because chlorine is used in a variety of ways, it is important to understand the human health and environmental impacts resulting from exposure.

### Human Effects

*Inhalation.* Acute exposure to chlorine gas results in pulmonary toxicity as well as irritation of the eyes, mucous membranes, and skin. There has been no clear correlation of permanent pulmonary damage caused by acute, high-concentration exposures to chlorine or chronic, low-concentration exposures to chlorine in the workplace, although individual differences do exist. There also do not appear to be any additional risks of cancer, reproductive effects, or developmental effects associated with chronic, low-concentration exposures nor with acute, high-concentration exposures in man or in animal models.

*Ingestion.* Chlorine is used in large quantities in the food processing industry to wash meat carcasses, seafood, vegetables, and fruits, and to improve flour qualities. There is no evidence to suggest that the ingestion of chlorine-processed food products is toxic to man.

*Environmental Effects.* Chlorine may impact the environment through the atmosphere, through precipitation, or may enter aquatic systems through the use of chlorine gas for disinfection and defouling of water and sewage effluents. The oxidation of natural and anthropogenic organics is the most important process in the decay of free chlorine in the environment. Chlorine is not known to bioaccumulate in the environment or the food chain; however, lipophilic oxidation products may potentially accumulate in animal tissues. Direct exposure of plants to chlorine gas results in a variety of symptom expression following toxic concentrations. There are no studies regarding the health effects of environmental exposures to atmospheric chlorine concentrations in wildlife or domesticated animals. Reports from catastrophic releases of chlorine have described the toxic effects of chlorine exposure to domesticated animals. In general, the clinical signs were due to the pulmonary toxicity of the chlorine gas.

*Chlorine in Water.*   Field and laboratory experiments have shown that chlorinated effluents are toxic to aquatic organisms. A chronic toxicity threshold for total residual chlorine has been calculated to be 0.0015 mg/L for freshwater organisms and 0.02 mg/L for marine organisms. Chlorine and chlorine reaction products are toxic to fish and invertebrates, although fish are generally less sensitive than invertebrate species.

## Regulatory Overview

*Chlorine Gas.*   The occupational exposure limits for chlorine in the air of work places vary in different countries. The ACGIH TLV for chlorine is 0.5 ppm with a STEL of 1 ppm. The OSHA PEL for chlorine is 1 ppm (CFR 1997a). The NRC, NIOSH, AIHA, and, most recently, the USEPA have derived short-term, one-time-only exposure levels during chemical emergency situations that are tiered based upon accepted level of toxic effect.

*Chlorine in Water.*   Chlorine is designated as a hazardous substance under Section 311(b)(2)(A) of the Federal Water Pollution Control Act and further regulated by the Clean Water Act Amendments of 1977 and 1978. Protective levels for freshwater and saltwater aquatic organisms have been set on the basis of total residual chlorine.

Under the provisions of the Safe Drinking Water Act (SDWA) Amendments of 1996, in 1998 EPA established a MRDL for chlorine of 4 mg/L.

## Acknowledgments

The author thanks The Chlorine Institute for the funding of this project and Dr. Kamal Singh for her insight and editorial review.

## Appendix

ACGIH   American Conference of Governmental Industrial Hygienists
AIHA     American Industrial Hygiene Association
AEGL    Acute Exposure Guideline Level

| BOD | Biochemical oxygen demand |
|---|---|
| CFR | Code of Federal Regulations |
| CIIT | Chemical Industry Institute of Toxicology |
| $CO_2$ | Carbon dioxide |
| D/DBP | Disinfectant/Disinfection Byproduct Rule |
| $EC_{50}$ | Effective concentration, 50%, that concentration necessary to achieve a 50% reduction in the desired effect (e.g., reproduction) |
| EEL | Emergency Exposure Limit |
| ERPG | Emergency Response Planning Guideline |
| FDA | Food and Drug Administration |
| $FEV_1$ | Forced Expiratory Volume in 1 sec |
| FIFRA | Federal Insecticide, Fungicide, and Rodenticide Act |
| FRC | Free residual chlorine; measurement of aqueous chlorine, includes $HOCl$ and/or $OCl^-$ |
| FVC | Forced vital capacity |
| HAP | Hazardous air pollutant |
| HCl | Hydrochloric acid |
| HMIS | Hazardous Material Identification System |
| HOCl | Hydrochlorous acid |
| HR | Hyperresponsiveness (airway) |
| IDLH | Immediately dangerous to life and health |
| $LC_{50}$ | Lethal concentration, 50%, that concentration which will kill half the experimental animals in an allotted time frame; a measure of acute toxicity. |
| MCL | Maximum contaminant level |
| MRDL | Maximum residual disinfectant level |
| NACE | National Association of Corrosion Engineers |
| NAPAP | National Acid Precipitation Assessment Program |
| NIOSH | National Institute for Occupational Safety & Health |
| NOAEL | No observed adverse effect level |
| NRC | National Research Council |
| O | Atomic oxygen |
| $O_3$ | Ozone |
| $OCl^-$ | Hypochlorite ion |
| OSHA | Occupational Safety and Health Administration |
| PEL | Permissible exposure level |
| ppbv | Parts-per-billion (volume) |
| ppm | Parts-per-million |
| RAD | Reactive airway disease |
| $RD_{50}$ | Respiratory rate depression, 50%, that concentration that will produce a 50% reduction in respiratory rate; a measure of sensory irritation. |
| RMP | Risk Management Program |
| SDWA | Safe Drinking Water Act |
| SPD | Spontaneous pulmonary disease |
| SPF | Specific pathogen free |

STEL      Short-term exposure limit
TLV       Threshold limit value
TSH       Total sulfhydryl content
TRC       Total residual chlorine; measurement of chlorine, includes FRC and
          chlorinated organics
TWA       Time-weighted average
USEPA     U.S. Environmental Protection Agency
UV        Ultraviolet

# References

Achon A, Roberts M (1996) Chlorine leak threatens Erkimia investment plans. Chem Week 158(4):24.

Adelson L, Kaufman J (1971) Fatal chlorine poisoning: report of two cases with clinico-pathologic correlation. Am J Clin Pathol 56(4):430–442.

Alarie Y (1973) Sensory irritation of the upper airways by airborne chemicals. Toxicol Appl Pharmacol 24:279–297.

Alderson R (1974) Sea-water chlorination and the survival and growth of the early developmental stages of plaice, *Pleuornectes platessa L.*, and Dover sole, *Solea solea* L. aquaculture 4:41–53.

ACGIH (American Conference of Governmental Industrial Hygienists) (1997) TLVs and BEIs. ACGIH, Cincinnati, OH.

American Industrial Hygiene Association (AIHA) (1996) Chlorine. In: Emergency Response Planning Guidelines. AIHA 211-EA-96 (last updated 1988). AIHA, Fairfax, VA.

Amoore JE, Hautala E (1983) Odor as an aid to chemical safety: odor thresholds compared with the threshold limit and volatiles for 214 industrial chemicals in air and water dilution. J Appl Toxicol 3:272–290.

Anglen DM (1981) Sensory Response of Human Subjects to Chlorine. Doctoral dissertation, University of Michigan, Ann Arbor, MI. (As cited in ACGIH 1991.)

Anonymous (1984) Chlorine Poisoning [Editorial] Lancet 1(8372):321–322.

Anonymous (1991) From the Centers of Disease Control: chlorine gas toxicity from mixture of bleach with other cleaning source. JAMA 266(18):2529.

Anonymous (1996) Leak warning. Chem Br 32(12):12.

Arthur JW, Eaton JG (1971) Chloramine toxicity to the amphipod (*Gammarus pseudolimnaeus*) and the fathead minnow (*Pimephales promelas*). J Fish Res Board Can 28:184.

Arthur JW, Andrew R, Mattson V, Olson D, Glass G, Halligan B, Walbridge C (1975) Comparative toxicity of sewage-effluent disinfection to freshwater aquatic life. Ecol Res Serv 60013-75-012 U.S. Environmental Protection Agency, Washington, DC.

Back KC, Thomas AA, MacEwen JC (1972) Reclassification of materials listed as transportation health hazards. Aerospace Medical Research Report TSA-20-72-3, Wright Patterson AFB, Ohio. (As cited in ACGIH 1991.)

Barbone F, Delzell E, Austin H, Cole P (1992) A case control study of lung cancer at a dye and resin manufacturing plant. Am J Ind Med 22:835–849.

Barregård, Sällsten G, Järvholm B (1990) Mortality and cancer incidence in chlor-alkali workers exposed to inorganic mercury. Br J Ind Med 47:99–104.

Barrow CS, Smith RG (1975) Chlorine-induced pulmonary function changes in rabbits. Am Ind Hyg Assoc J 36:398–403.

Barrow CS, Steinhagen WH (1982) Sensory irritation tolerance development to chlorine in F-344 rats following repeated inhalation. Toxicol Appl Pharmacol 65:383–389.

Barrow CS, Alaire Y, Warrick JC, Stock MF (1977) Comparison of the sensory irritation response in mice to chlorine and hydrogen chloride. Arch Environ Health 32:68–76.

Barrow CS, Kociba RJ, Rampy LW, Keyes DG, Albee RR (1979) An inhalation toxicity study of chlorine in Fischer-344 rats following 30 days of exposure. Toxicol Appl Pharmacol 49:77–88.

Basch RC, Truchan JG (1976) Toxicity of chlorinated power plant condenser cooling waters to fish. Ecological Research Series. EPA-600/3-76-009. Office of Research and Development, Environmental Research Laboratory, U.S. Environmental Protection Agency, Duluth, MN.

Bass ML, Heath AG (1977) Toxicity of intermittent chlorination to bluegill, *Lepomis macrochirus*: interaction with temperature. Bull Environ Contam Toxicol 17(4):416–423.

Beck H (1959) Experimental Determination of the Olfactory Thresholds of Some Important Irritant Gases (Chlorine, Sulfur Dioxide, Ozone, Nitrous Gases) and Symptoms Induced in Humans by Low Concentrations. Doctoral dissertation. Julius-Maximilians-Universitat, Wurzberg. (As cited in NIOSH 1976.)

Bell DP, Elmes PC (1965) The effects of chlorine gas on the lungs of rats without spontaneous pulmonary disease. J Pathol Bacteriol 89:307–317.

Bellanca MA, Bailey DS (1977) Effects of chlorinated effluents on aquatic ecosystem in the lower James River. J Water Pollut Control Fed 49(4):639–645.

Benedict HM, Breen WH (1955). The use of weeds as a means of evaluating vegetation damage caused by air pollution. In: Proceedings of the Third National Air Pollution Symposium, Pasadena, CA, pp 177–190.

Bennet JH, Hill AC (1975) Interactions of air pollutants with vegetation. In: Mudd JB, Kozlowski TT (eds) Responses of Plants to Air Pollution. Academic Press, New York, p 277.

Bherer L, Cushman R, Courteau JP, Quevillon M, Cote G, Bourbeau J, L'Archeveque J, Cartier A, Malo JL (1994) Survey of construction workers repeatedly exposed to chlorine over a three to six-month period in a pulpmill: II. Follow up of affected workers by questionnaire, spirometry, and assessment of bronchial responsiveness 18 to 24 months after exposure ended. Occup Environ Med 51(4):225–228.

Bloomfield AE (1959) Domestic chlorine poisoning. Br Med J 2:1332.

Bond GG, Wight PC, Flores GH, Cook RR (1983) A case-control study of brain tumor mortality at a Texas chemical plant. J Occup Med 25(5):377–386.

Bond GG, Shellenberger RJ, Flores GH, Cook RR, Fishbeck WA (1985) A case-control study of renal cancer mortality at a Texas chemical plant. Am J Ind Med 7(2):123–139.

Bond GG, Flores GH, Shellenberger RJ, Cartmill JB, Fishbeck WA, Cook RR (1986) Nested case-control study of lung cancer among chemical workers. Am J Epidemiol 124:53–66.

Booth PM, Sellers CM, Garrison NE (1981) Effects of intermittent chlorination on plasma proteins of rainbow trout (*Salmo gairdneri*). Bull Environ Contam Toxicol 26(2):163–170.

Boulet LP (1988) Increases in airway responsiveness following acute chlorine exposure to respiratory irritants. Chest 94:476–481.

Brennan E, Leone IA, Daines RH (1965) Chlorine as a phytotoxic air pollutant. Air Water Pollut 9(12):791–797.

Brennan E, Leone IA, Daines RH (1966) Response of pine trees to chlorine in atmosphere. For Sci 12:386–390.

Brennan E, Leone IA, Holmes C (1969) Accidental chlorine gas damage to vegetation. Plant Dis Rep 53:875–878.

Brooks AJ, Baker AL (1972) Chlorination of power plants: impact on phytoplankton productivity. Science 176:1414–1415.

Brooks AS, Seegert GL (1977) The effects of intermittent chlorination on rainbow trout and yellow perch. Trans Am Fish Soc 106:278–286.

Brooks AS, Bartos JM (1984) Effects of free and combined chlorine and exposure duration on rainbow trout, channel catfish, and emerald shiners. Trans Am Fish Soc 113: 786–93.

Buckley LA, Jiang XZ, James RA, Morgan KT, Barrow CS (1984) Respiratory tract lesions induced by sensory irritants at the $RD_{50}$ concentration. Toxicol Appl Pharmacol 74:417–429.

Budaveri S (1996) In: Budavari S, O'Neil MJ, Smith A, Heckelman PE, Kinnieary JF (eds) The Merck Index, 12th Ed. Merck, Whitehouse Station, NJ.

Burton DT, Hall LW Jr, Margrey SL, Small RD (1979) Interactions of chlorine, temperature change ($\Delta$T) and exposure time on survival of striped bass (*Morone saxatilis*) eggs and prolarvae. J Fish Res Board Can 36:1108–1113.

Capuzzo JM (1979) The effects of halogen toxicants on survival, feeding and egg production of the rotifer *Brachionus plicatilis*. Estuar Coast Mar Sci 8:307–316.

Capuzzo JM, Lawrence SA, Davidson JA (1976) Combined toxicity of free chlorine, chloramine and temperature to stage I larvae of the American lobster, *Homarus americanus*. Water Res 10(12):1093–1099.

Capuzzo, JM, Davidson JA, Lawrence SA, Ubni M (1977) The differential effects of free and combined chlorine on juvenile marine fish. Estuarine Coastal Mar Sci 5: 733–741.

Carlton BD, Bartlett P, Basaran A, Colling K, Osis I, Smith MK (1986) Reproductive effects of alternate disinfectants. Environ Health Perspect 69:237–241.

Carpenter EJ, Peck BB, Anderson SJ (1972) Cooling water chlorination and productivity of entrained phytoplankton. Mar Biol 16:37–40.

Chang JCF, Barrow CS (1984) Sensory irritation tolerance and cross-tolerance in F-344 rats exposed to chlorine or formaldehyde gas. Toxicol Appl Pharmacol 76(2): 319–327.

Charan, NB, Lakshminarayan S, Myers GC, Smith DD (1985) Effects of accidental chlorine inhalation on pulmonary function. West J Med 143:333–336.

Chasis H, Zapp JA, Bannon JH, Whittenberger JL, Helm J, Doheny JJ, MacLeod CM (1947) Chlorine accident in Brooklyn. Occup Med 4(2):152–176. (As cited in NIOSH 1976.)

Chemical Marketing Reporter (1989) Chemical profile: Chlorine. June 12. Vol 235 Iss 24.

Chemical Marketing Reporter (1995) Chemical profile: Chlorine. June 12, 1995. Vol 247 Iss 24.

Cherry DS, Hoehn RC, Waldo SS, Willis DH, Cairns J Jr, Dickson KL (1977) Field-laboratory determined avoidance of the spotfin shiner (*Notropis spilopterus*) and the bluntnose minnow (*Pimephales notatus*) to chlorinated discharge. Water Res Bull 13: 1047–1055.

Cherry DS, Larrick SR, Giattina JD, Cairns J Jr, Van Hassel J (1982) Influence of temperature selection upon the chlorine avoidance of cold-water and warm-water Fishes. Can J Fish Aquat Sci 39:162–173.

Chester EH, Gillespie DG, Krause FD (1969) The prevalence of chronic obstructive pulmonary disease in chlorine gas workers. Am Rev Respir Dis 99:365–373.

Chester EH, Kaimal J, Payne CB Jr, Kohn PM (1977) Pulmonary injury following exposure to chlorine gas. Possible beneficial effects of steroid treatment. Chest 72(2): 247–250.

Chlorine Institute (1990a) Pamphlet 90: Toxicity summary for chlorine and hypochlorites, and chlorine in drinking water. Chlorine Institute, Washington, DC.

Chlorine Institute (1990b) Pamphlet 84: Environmental fate of chlorine in the atmosphere. Chlorine Institute, Washington, DC.

Chlorine Institute (2000) Pamphlet 10: North American chlor-alkali industry. Plants and production data book. Chlorine Institute, Washington, DC.

Clarke EGC, Clarke ML (eds) (1981) Veterinary Toxicology. Bailliere-Tindall, London, p 82.

Code of Federal Regulations (CFR) (1997a) Labor. 29 Part 1910. §1910.1000.

Code of Federal Regulations (CFR) (1997b) Protection of environment. 40 Part 180, Subpart D. §180.1095.

Code of Federal Regulations (CFR) (1997c) Food and drugs. 21 Part 137.

Code of Federal Regulations (CFR) (1998) Protection of environment. 40 Part 141 §141.54.

Colardyn F, van der Straeten M, Tasson J, van Egmond J (1976) Acute chlorine gas intoxication. Acta Clin Belg 31(2):70–77.

Collins JS (1958) Some experiences with Nais and nematodes in the public water supply of Norwich. Proc Soc Water Treat Exam 7:157.

Cordasco EM, Gregory R, Popovici M, Goodrich JL, Morgat EH, Mallinchok J, Gideon E, Van Ordstrand HS, Mosher WE, del Greco F (1977) The health effects of halogens in air pollution. Occup Health Saf 46(1):36–38.

Coughlan J, Davis MH (1981) Effect of chlorination on plankton at several United Kingdom coastal power stations. In Jolley RL, Brungs WA, Cotruvo JA, Cumming RB, Mattice JS, Jacobs VA (eds) Water Chlorination Environmental Impact and Health Effects, Vol. 4. Ann Arbor Science Publishers, Ann Arbor, MI, pp 153–1066.

Cunningham HM, Lawrence GA (1978) Effect of chlorinated lipid and protein fractions on growth rate and organ weights of rats. Bull Environ Contam Toxicol 19:73–79.

Cunningham HM, Lawrence GA, Tryphonas L (1977) Toxic effects of chlorinated cake flour in rats. J Toxicol Environ Health 2(5):1161–1171.

D'Alessandro A, Kuschner W, Wong H, Boushey HA, Blanc PD (1996) Exaggerated responses to chlorine inhalation among persons with non-specific airway hyperactivity. Chest 109(2):331–337.

Decker WJ, Koch HF (1978) Chlorine at the swimming pool: an overlooked hazard. Clin Toxicol 13(3):377–381.

Deschamps D, Soler P, Rosenberg N, Baud F, Gervais P (1994) Persistent asthma after inhalation of a mixture of sodium hypochlorite and hydrochloric acid. Chest 105(6): 1895–1896.

Dewhurst F (1981) Voluntary chlorine inhalation [Letter]. Br Med J Clin Res Ed 282(6263):565–566.

Dickson KL, Cairns JR Jr, Gregg BC, Messenger DI, Plafkin JL, van der Schalie WH

(1977) Effects of intermittent chlorination on aquatic organisms and communities. J Water Pollut Control Fed 49:35–44.

Dinnel PA, Stover QJ, DiJulio DH (1979) Behavioral responses of shiner perch to chlorinated primary sewage effluent. Bull Environ Contam Toxicol 22(4–5):708–714.

Dodd DE, Bus JS, Barrow CS (1980) Lung sulfhydryl changes in rats following chlorine inhalation. Toxicol Appl Pharmacol 52:199–208.

Donnelly SC, Fitzgerald MX (1990) Reactive airways dysfunction syndrome (RADS) due to chlorine gas exposure. Ir J Med Sci 159:275–277.

Dotson DA, Helz GR, Sugam R (1986) Mineralization of organic matter and other chemical effects of chlorination. Water Res 20:1031–1039.

Downs AJ, Adams CJ (1973) Chlorine, bromine, iodine and astatine. In: Bailar JC, Emeleús HJ, Nyholm, R and Trotman-Dickenson AF Comprehensive Inorganic Chemistry, Vol. 2. Pergamon Press, Oxford, pp 1107–1594 Chpt 26.

Duce RA (1969) On the source of gaseous chlorine in the marine atmosphere. J Geophys Res 74:4597–4599.

Druckery H (1968) Chlorinated drinking water, toxicity tests, involving seven generations of rats (in German). Food Cosmet Toxicol 6:147–154. (As cited in IARC Monographs, Vol. 52.)

Ellenhorn MJ, Barceloux DG (eds) (1988) Chlorine: airborne toxins. In Medical Toxicology: Diagnosis and Treatment of Human Poisoning. Elsevier, New York, pp 878–879.

Ellis MM (1937) Detection and measurement of stream pollution. Bull Bur Fish 48:365.

Elmes PC, Bell D (1963) The effects of chlorine gas on the lungs of rats with spontaneous pulmonary disease. J Pathol Bacteriol 86:317–326.

Faigel HC (1964) Hazards to health: mixtures of household cleaning agents. N Engl J Med 271:618.

Fandrei G, Collins HL (1979) Total residual chlorine: the effect of short-term exposure on the emerald shiner Notropis atherinoides (Rafinesque). Bull Environ Contam Toxicol 23:262–268.

Ferris BG, Burgess WA, Worchester J (1967) Prevalence of chronic respiratory disease in a pulpmill and paper mill in the United States. Br J Ind Med 24:26–37.

Ferris BG, Puleo S, Chen HY (1979) Mortality and morbidity in a pulp and a paper mill in the United States: a ten-year follow-up. Br J Ind Med 36:127–134.

Fieldner AC, Katz SH, Kinney SP (1921) Gas masks for gasses met in fighting fires. Bureau of Mines Tech Pap 248. U.S. Dept of the Interior, Bureau of Mines, Washington, DC, pp 3–61. (As cited in NIOSH 1976.)

Fisher N, Berry R, Hardy J (1983a) Short-term study in rats of chlorinated cake flour. Food Chem Toxicol 21(4):423–426.

Fisher N, Hutchinson JB, Berry R, Hardy J, Ginocchio AV (1983b) Long-term toxicity and carcinogenicity studies of cake made from chlorinated flour. 1. Studies in Rats. Food Chem Toxicol 21(4):427–434.

Fleta J, Calvo C, Zuniga J, Castellano M, Bueno M (1986) Intoxication of 76 children by chlorine gas. Hum Toxicol 5(2):99–100.

Frank R (1986) Acute and chronic respiratory effects of exposure to inhaled toxic agents. In: Merchant JA (ed) Occupational Respiratory Diseases, pp 571–83. NIOSH Pub 82–102. NIOSH, Washington, DC.

Gapany-Gapanavicius M, Yellin A, Almog S, Tirosh M (1982) Pneumomediastinum. A complication of chlorine exposure from mixing household cleaning agents. JAMA 248(3):349–350.

Giauque WF, Powell TM (1939) Chlorine. The heat capacity, vapor pressure, heats of fusion and vaporization and entropy. J Am Chem Soc 61:1970–1974.

Ginocchio AV, Fisher N, Hutchinson JB, Berry R, Hardy J (1983) Long-term toxicity and carcinogenicity studies of cake made from chlorinated flour. 2. Studies in mice. Food Chem Toxicol 21(4):435–439.

Givan DC, Eigen H, Tepper RS (1989) Longitudinal evaluation of pulmonary function in an infant following chlorine gas exposure. Pediatr Pulmonol 6:191–194.

Graedel TE, McGill R (1986) Degradation of materials in the atmosphere. Environ Sci Technol 20:1093–1100.

Gregg BC (1974) The effects of chlorine and heat on selected stream invertebrates. Ph.D. Thesis. Virginia Polytechnic Institute and State University, Blacksburg.

Hall LW Jr, Margrey SL, Graves WC, Burton DT (1981) Avoidance responses of juvenile Atlantic menhaden, *Brevoortia tyrannus*, subjected to simultaneous chlorine and ΔT conditions. In: Jolley RL, Brungs WA, Cotruvo JA, Cumming RB, Mattice JS, Jacobs VA (eds) Water Chlorination: Environmental Impact and Health Effects, Vol. 4. Ann Arbor Science Publishers, Ann Arbor, MI, pp 983–991.

Hall LW Jr, Burton DT, Margrey SL, Graves WC (1982a) A comparison of the avoidance of individual and schooling juvenile Atlantic menhaden, *Brevoortia tyrannus* subjected to simultaneous chlorine and delta T conditions. J Toxicol Environ Health 10(6):1017–1026.

Hall LW Jr, Graves WC, Burton DT, Margrey SL, Hetrick FM, Roberson BS (1982b) A comparison of chloride toxicity to three life stages of striped bass (*Morone saxatilis*). Bull Environ Contam Toxicol 29(6):631–636.

Hammer MJ (1975) Water Processing: Chlorination. In: Water and Waste-Water Technology pp 238–242 Chpt 7 John Wiley & Sons, Inc. New York.

Harper DS, Jones RD (1982) The relative sensitivity of fifty plant species to chronic doses of hydrogen chloride. Phytopathology 72:261–262.

Heidemann SM, Goetting MG (1991) Treatment of acute hypoxemic respiratory failure caused by chlorine exposure. Pediatr Emerg Care 7(2):87–88.

Heldaas SS, Langård S, Anderson A (1989) Incidence of cancer in a cohort of magnesium production workers. Br J Ind Med 46:617–623.

Hirsch AR (1995) Chronic neurotoxicity of acute chlorine gas exposure. Neurotoxicology 19(4):760.

Hinrichsen D (1987) The forest decline enigma. What underlies extensive dieback on two continents? Bioscience 37(1):542–546.

Hopp V (1981) On organohalogens? Chemiker-Zeitung 115:34. (As cited in Munro 1994.)

Hose JE, Stoffel RJ (1980) Avoidance response of juvenile *Chromis punctipinnis* to chlorinated seawater. Bull Environ Contam Toxicol 25(6):929–935.

Hoss DE, Coston LC, Baptist JP, Engel DW (1975) Effects of temperature, copper and chlorine on fish during simulated entrainment in power-plant condenser cooling systems. In: Environmental Effects of Cooling Systems at Nuclear Power Plants. IAEA-SM-187/19. Int Atom Energy Agency, Vienna, pp 519–527.

IARC (1991) Chlorinated Drinking-Water; Chlorination By-Products: Some Other Halogenated Compounds; Cobalt and Cobalt Compounds. IARC Monographs on the Evaluation of Carcinogenic Risks to Humans, Vol. 52. World Health Organization International Agency for Research on Cancer, Lyon, France.

Jäppinen P, Hakulinen T, Pukkala E, Toal S, Kurppa K (1987) Cancer incidence of workers in the Finnish pulp and paper industry. Scand J Work Environ Health 13: 197–202.

Jiang XZ, Buckley LA, Morgan KT (1983) Pathology of toxic responses to the $RD_{50}$ concentration of chlorine gas in the nasal passages of rats and mice. Toxicol Appl Pharmacol 71(2):225–236.

Jolley RL (1984) Basic issues in water chlorination: a chemical perspective. In: Jolley RL, Bull RJ, Davis WP, Katz S, Roberts MJ, Jacobs VA (eds) Water Chlorination Chemistry, Environmental Impact and Health Effects, Vol. 5. Lewis Publishers, Chelsea, MI. pp 19–38.

Jolley RL, Carpenter JH (1984) A review of the chemistry and environmental fate of reactive oxidant species in chlorinated water. In: Jolley RL, Bull RJ, Davis WP, Katz S, Roberts MJ, Jacobs VA (eds) Water Chlorination: Environmental Impact and Health Effects, Vol. 5. Lewis Publishers, Chelsea, MI., pp 3–47.

Johnson JD, Jensen JN (1986) THM and TOX formation: routes, rates and precursors. J AWWA 78(4): 156–162.

Jones FL (1972) Chlorine poisoning from mixing household cleaners. JAMA 222(10): 1312.

Jones RN, Hughes JM, Glindmeyer H, Weill H (1986) Lung function after acute chlorine exposure. Am Rev Respir Dis 134(6):1190–1195.

Joyner RE, Durel EG (1962) Accidental liquid chlorine spill in a rural community. J Occup Med 4(3):152–154.

Kaufman J, Burkons D (1971) Clinical, roentgenologic, and physiologic effects of acute chlorine Exposure. Arch Environ Health 23(1):29–34.

Key PB, Scott GI (1986) Lethal and sublethal effects of chlorine, phenol, and chlorine-phenol mixtures on the mud crab, *Panopeus herbstii*. Environ Health Perspect 69: 307–312.

Kilburn KH (1995) Evidence that inhaled chlorine is neurotoxic and causes airway obstruction. Int J Occup Med Toxicol 4(2):267–276.

Klonne DR, Ulrich E, Riley GM, Hamm TE Jr, Morgan KT, Barrow CS (1987) One-year inhalation toxicity study of chlorine in rhesus monkeys (*Macaca mulatta*). Fundam Appl Toxicol 9:557–572.

Kotula AW, Emswiler-Rose BS, Cramer DV (1987) Subacute study of rats fed ground beef treated with aqueous chlorine: hematologic and clinical pathology. J Toxicol Environ Health 20(4):401–409.

Kulp K, Tsen CC, Daly CJ (1972) Effect of chlorine on starch component of soft-wheat flour. Cereal Chem 49:194–200.

Larson GL, Warren CE, Hutchina FE, Lampert LP, Schlesinger DA, Seim WK (1978) Toxicity of residual chlorine compounds to aquatic organisms. EPA-600/3-78-023. U.S. Environmental Protection Agency, Washington, DC.

Laubusch EJ (1962a) Physical and Chemical Effects of Chlorine. In: Sconce JS (ed) Chlorine. Its Manufacture, Properties, and Uses. American Chemical Society Monograph Series No. 154. Reinhold, New York, pp 21–45. Chpt. 3.

Laubusch EJ (1962b) Waste Water Chlorination. In: Sconce JS (ed) Chlorine. Its Manufacture, Properties, and Uses. American Chemical Society Monograph Series no. 154. Reinhold, New York, pp 485–511. Chpt. 15.

Laubusch EJ (1962c) Water Chlorination. In: Sconce JS (ed) Chlorine. Its Manufacture, Properties, and Uses. American Chemical Society Monograph Series no. 154. Reinhold, New York, pp 457–484. Chpt. 14.

Learner MA, Edwards RW (1963) The toxicity of some substances to *Nais* (Oligochaeta). Proc Soc Water Treat Exam 12:161.

Leininger JR, Jarabek AM, Morgan KT (1994) Comparison of nasal mucous cell response following chronic inhalation of chlorine in rodents and monkeys. Toxicologist 14:314.

Lemiere C, Malo J-L, Boutet M (1997) Reactive airways dysfunction syndrome due to chlorine: sequential bronchial biopsies and functional assessment. Eur Respir J 10(1): 241–244.

Leonardos G, Kendall D, Barnard NJ (1968) Odor threshold determination of 53 odorant chemicals. Presented at 61st Annual Meeting, Air Pollution Control Association (Paper 68-13), St. Paul, MN, June 23–27. (As cited in NIOSH 1976.)

Levy JM, Hessel SJ, Nykamp PW, Stegman CJ, Crowe JK, Spiegel RM, Horsley WW, Cook GC (1986) Detection of the cerebral lesions of chlorine intoxication by magnetic resonance imaging. Magn Reson Imaging 4(1):51–52.

Lide DR (ed) (1998) CRC Handbook of Chemistry and Physics. A Ready-Reference Book of Chemical and Physical Data, 78th Ed. CRC Press, New York.

Loewenstein LM, Anderson JG (1984) Rate and product measurements for the reactions of OH with $Cl_2$, $Br_2$, and BrCl at 298 K. J Phys Chem 88:6277–6286.

MacEwen JD, Vernot EH (1972) Toxic Hazards Research Unit Annual Technical Report. AMRL-TR-72-62. Aerospace Medical Research Laboratory, Wright-Patterson Air Force Base, OH. NTIS, Springfield, VA. (As cited in USEPA 1997.)

Malone LJSD, Warin JF (1945) A domestic case of chlorine-gas poisoning. Br Med J 1: 14.

Martinez T, Long C (1995) Explosion risk from swimming pool chlorinators and review of chlorine toxicity. J Toxicol Clin Toxicol 33(4):349–354.

Masri MS (1986) Chlorinating poultry chiller water: the generation of mutagens and water re-use. Food Chem Toxicol 24(9):923–930.

Matthews RC, Bosnak AD, Tennant DS, Morgan DS, Morgan EL (1977) Mortality curves of blind cave crayfish (*Orconectes australis*) exposed to chlorinated stream water. Hydrobiologia 53(2):107–111.

Mattice JS, Zittel HE (1976) Site-specific evaluation of power plant chlorination. J Water Poll Control Fed 48:2284–2308.

Mattice JS, Burch MB, Tsai SC, Roy WK (1981) A toxicity testing system for exposing small invertebrates and fish to short square-wave concentrations of chlorine. Water Res 15:923–927.

McConnel JC, Henderson GS, Barrie L, Bottenheim J, Niki H, Langford CH, Templeton EMJ (1992) Photochemical bromine production implicated in Arctic boundary-layer ozone depletion. Nature (Lond) 355(6356):150–152.

McElroy MB, Salawitch RJ (1989) Changing composition of global stratosphere. Science 2463:763–770.

McNulty MJ, Chang JCF, Barrow CS, Casanova-Schmitz M, d'A Heck H (1983) Sulfhydryl oxidation in rat nasal mucosal tissues after chlorine inhalation. Toxicol Lett 17: 241–246.

Meakins JC (1919) The after-effects of chlorine gas poisoning. Can Med Assoc J 9: 968–974.

Middaugh DP, Crane AM, Couch JA (1977a) Toxicity of chlorine to juvenile spot, *Leiostomus xanthurus*. Water Res 11:1089–1096.

Middaugh DP, Couch JA, Crane AM (1977b) Responses of early life history stages of the striped bass, *Morone saxatilis*, to chlorination. Chesapeake Sci 18:141–153.

Morgan RP, Prince RD (1977) Chlorine toxicity to eggs and larvae of five Chesapeake Bay fishes. Trans Am Fish Soc 106:380–385.

Mrvos R, Dean BS, Krenzelok EP (1993) Home exposures to chlorine/chloramine gas: review of 216 cases. South Med J 86(6):654–657.

Muchmore D, Eppel D (1973) The effects of chlorination of wastewater on fertilization in some marine invertebrates. Mar Biol 19:93–95.

Munro I (1994) Interpretive review of the potential adverse effects of chlorinated organic chemicals on human health and the environment. Regul Toxicol Pharmacol 20(1): S69–S108.

Mustchin CP, Pickering CA (1979) "Coughing water": bronchial hyperreactivity induced by swimming in a chlorinated pool. Thorax 34(5):682–683.

NACE (1984) Corrosion Basics. National Association of Corrosion Engineers, Houston, TX.

NAPAP (1987) Interim Assessment: The Causes and Effects of Acid Deposition, Vol. 4. National Acid Precipitation Assessment Program, Washington, DC. (As cited in Munro 1994.)

NAPIM (1984) HMIS: Guidelines for a Hazardous Materials Identification System for Raw Materials. National Association of Printing Ink Manufacturers, Harrison, NY.

NAS (1976) Chlorine and Hydrogen Chloride. National Academy of Science, Washington, DC.

NAS (1984) Emergency and Continuous Exposure Guidance Levels for Selected Airborne Contaminants, Vol. 2. National Research Council Committee on Toxicology. National Academy Press, Washington, DC.

NIOSH (1976) Criteria of a recommended standard: occupational exposure to chlorine. National Institute for Occupational Safety and Health (NIOSH). U.S. Department of Health, Education and Welfare, Public Health Service, Washington, DC.

NIOSH (1987) NIOSH respirator decision logic. DHHS (NIOSH) Publication 87-108. NIOSH, Washington, DC.

NIOSH (1994) Pocket guide to chemical hazards. National Institute for Occupational Safety and Health (NIOSH). U.S. Dept. HEW, Public Health Service, Washington, DC.

Nodelman V, Ultman JS (1999a) Longitudinal distribution of chlorine absorption in human airways: comparison of nasal and oral quiet breathing. J Appl Physiol 86:1984–1993.

Nodelman V, Ultman JS (1999b) Longitudinal distribution of chlorine absorption in human airways: a comparison to ozone absorption. J Appl Physiol 87(6):2073–2080.

Nodelman V, Ben-Jebria A, Ultman JS (1998) Fast-responding thermionic chlorine analyzer for respiratory applications. Rev Sci Instrum 69(11):3978–3983.

Ogundele GI (1989) The effect of cation on the corrosion of carbon and stainless steels in differing chloride environments. Corrosion 45:981–982.

Osbourn LL (1984) Response of Sheep River, Alberta, macroinvertebrate communities to discharge of chlorinated municipal sewage effluent. In: Jolley RL, Bull RJ, Davis WP, Katz S, Roberts MJ, Jacobs VA (eds) Water Chlorination Chemistry, Environmental Impact and Health Effects, Vol. 5. Lewis Publishers, Chelsea, MI. pp 481–492.

Owusu-Yaw J, Wheeler WB, Wei CI (1991) Genotoxicity studies of the reaction of chlorine or chlorine dioxide with L-tryptophan. Toxicol Lett 56(1-2):213–227.

Patil LRS, Smith RG, Vorwald AJ, Mooney TF (1970) The health of diaphragm cell workers exposed to chlorine. Am J Ind Hyg 31:678–686.

Patterson JT (1968) Bacterial flora of chicken carcasses treated with high concentrations of chlorine. J Appl Bacteriol 31(4):544–550.

Penington AH (1954) War gases and chronic lung disease. Med J Aust 1:510.

Phillip R, Shepherd C, Fawthrop F, Poulsom B (1985) Domestic chlorine poisoning [Letter]. Lancet 2(8453):495.

Polysongsang Y, Beach BC, Dilisio RE (1982) Pulmonary function changes after acute inhalation of chlorine gas. South Med J 75(1):23–26.

Pratt J, Bowers NJ, Niederlander BR, Calms J Jr (1988) Effects of chlorine on microbial communities in naturally derived microcosms. Environ Toxicol Chem 7:679–687.

Puente ME, Vega-Villasante F, Holguin G, Bashan Y (1992) Susceptibility of the brine shrimp *Artemia* and its pathogen *Vibrio parahaemolyticus* to chlorine dioxide in contaminated seawater. J Appl Bacteriol 73(6):465–471.

Rafferty P (1980) Voluntary chlorine inhalation: a new form of self-abuse? Br Med J 281:1178–1179.

Roberts MH, Diaz RJ, Bender ME, Huggett RJ (1975) Acute toxicity of chlorine to selected estuarine species. J Fish Res Board Can 32:2525–2528.

Rotman HH, Fliegelman MJ, Moore T, Smith RG, Anglen DM, Kowalski CJ, Weg JG (1983) Effects of low concentrations of chlorine on pulmonary function in humans. J Appl Physiol 54:1120–1124.

RTECS (1997) Chlorine record. National Library of Medicine, Toxicology Information Program On-line Services. Occupational exposure level (OEL) fields; updated January 1993.

Rupp H, Henschler D (1967) Effects of low chlorine and bromide concentrations in man. Arch Gewerbepathol Gewerbehyg 23:79–90. (As cited in NIOSH 1976.)

Ruth JH (1986) Odor thresholds and irritation levels of several chemical substances: a review. Am Ind Hyg Assoc J 47:A142–A151.

Ryazanov VA (1962) Sensory physiology as basis for air quality standards. Arch Environ Health 5:480–489.

Sandall TE (1922) The later effects of gas poisoning. Lancet 2:857–859.

Schlagbauer M, Henschler D (1967) Toxicity of chlorine and bromine after single and repeated inhalation. Arch Gewerbepathol and Gewerbehyg 23:91–98. (As cited in WHO 1982.)

Schroff CP, Khade MV, Srinivasan M (1988) Respiratory cytopathology in chlorine gas toxicity: a study in 28 subjects. Diagn Cytopathol 4(1):28–32.

Schumacher PD, Ney JJ (1980) Avoidance response of rainbow trout (*Salmo gairdneri*) to single-dose chlorination in a power plant discharge canal. Water Res. 14:651–655.

Schwartz DA, Smith DD, Lakshminarayan S (1990) The pulmonary sequelae associated with accidental inhalation of chlorine gas. Chest 97:820–825.

Sconce J (ed) (1962) Chlorine. Its Manufacture, Properties, and Uses. American Chemical Society Monograph Series 154. Reinhold, New York.

Seegert GL, Brooks AS (1978) The effects of intermittent chlorination on coho salmon, alewife, spottail shiner, and rainbow smelt. Trans Am Fish Soc 107:346–353.

Seegert GL, Brooks AS, Latimer DL (1977) The effects of a 30-minute exposure of selected Late Michigan fishes and invertebrates to residual chlorine. In: Jensen LD (ed) Biofouling and Control Procedures: Technology and Ecological Effects. Dekker, New York, pp 91–99.

Silver SD, McGrath FP (1942) Chlorine. Median lethal concentration for mice. Edgewood Arsenal Technical Reports 351, 373. War Department, Chemical Warfare Service, Edgewood Arsenal, MD. (As cited in AIHA 1996.)

Singh HB, Kasting JF (1988) Chlorine-hydrocarbon photochemistry in the marine troposphere and lower stratosphere. J Atmos Chem 7:261–285.

Sklyanskaya RM, Rappoport TL (1935) Experimentelle Studien Über Chronische Vergiftung Von Kaninchen Mit Geringen Chlorkonzentrationen und Die Entwicklung Der Nachkommenschaft Der Chlorverigifeten Kaninchen (in German). Arch Exp Pathol Pharmakol 117:276–285. (As cited in WHO 1982.)

Smith MK, Habash DL, Colling KA, Basaran AH, Osis ID, Carlton BD. (1985) Examination of Potential Reproductive Effects of Chlorine Administered to Long-Evans Rats by Gavage. (abstract). The Toxicologist. 5(1):185.

Stevens AA, Moore L, Dressman RC, Seeger DR (1985) Disinfection chemistry in drinking water: overview of impacts on drinking water chemistry. In: Safe Drinking Water: The Impacts of Chemicals on a Limited Source. RG Rice (ed) Drinking water Research Foundation, Chelsea, MI, Lewis Publishers, pp 87–108. (As cited in Chlorine Institute 1990a.)

Stober QJ, Hanson CH (1974) Toxicity of chlorine and heat to pink, *Onchorhynchus gorbuscha*, and Chinook salmon, *Onchorhynchus tshawytscha*. Trans Am Fish Soc 103:569–576.

Stober QJ, Dinnel PA, Hurlburt EF, DiJulio DH (1980) Acute toxicity and behavioral responses of coho salmon (*Oncorhynchus kisutch*) and shiner perch (*Cymatogaster aggregate*) to chlorine in heated seawater. Water Res 14:347–354.

Stokinger HE (1981) Chlorine. In: Clayton G (ed) Patty's Industrial Hygiene and Toxicology, Vol. 2b. Wiley, New York, pp 2954–2959.

Symonds RB, Rose WI, Reed MH (1988) Contribution of Cl-, F-fearing gases to the atmosphere by volcanoes. Nature (Lond) 334(6181):415–418.

Szerlip HM, Singer I (1984) Hyperchloremic metabolic acidosis after chlorine inhalation. Am J Med 77(3):581–582.

Temple WA, Dobbinson TL (1983) Acute chlorine poisoning from a high schools experiment. N Z Med J 96(740):720–721.

Thatcher TO, Schneider MJ, Wolf EG (1976) Bioassays on the combined effects of chlorine, heavy metals and temperature on fishes and fish food organisms. Part I. Effects of chlorine and temperature on juvenile brook trout (*Salvelinus fontinolis*). Bull Environ Contam Toxicol 15(1):40–48.

Tsai C (1968) Effects of chlorinated sewage effluents on fishes in upper Patuxent River, Maryland. Chesapeake Sci 9:83–92.

Tsai C (1970) Changes in fish populations and migrations in relation to increased sewage pollution in Little Patuxent River, Maryland. Chesapeake Sci 11:34–41.

Tsai C (1973) Water quality and fish life below sewage outfalls. Trans Am Fish Soc 102(2):281–291.

Tsen CC, Kulp K (1971) Effects of chlorine in flour proteins, dough properties, and cake quality. Cereal Chem 48:247–253.

Turner A, Chu A (1981) Chlorine toxicity as a function of environmental variables and species tolerance. In: Jolley RL, Brungs WA, Cotruvo JA, Cumming RB, Mattice JS, Jacobs VA (eds) Water Chlorination Environmental Impact and Health Effects, Vol. 4. Ann Arbor Science Publishers, Ann Arbor, MI, pp 927–946.

Underhill FP (1920) The Lethal War Gasses: Physiology and Experimental Treatment. Yale University Press, New Haven, pp 3–10. (As cited in AIHA 1996.)

USEPA (1974) Conference on Recycling Treated Municipal Wastewater Through Forest Cropland. EPA-660/2-74-003. Office of Research and Development, U.S. Environmental Protection Agency, Washington, DC.

USEPA (1977) Water requirements for steam-electric power generation and synthetic fuel plants in the western United States. EPA-600/7-77-037. Office of Research and Development, USEPA, Washington, DC.

USEPA (1985) Ambient water quality criteria for chlorine. EPA-440/5-84-030. USEPA, Washington, DC.

USEPA (1995) Toxic release inventory. TRI database. Office of Pollution Prevention and Toxics, USEPA, Washington, DC.

USEPA (1997) National Advisory Committee for Acute Exposure Guideline Levels for Hazardous Substances. Federal Register Notice. 62FR 58840. October 30, 1997.

Venkataramiah A, Lakshi GJ, Best CM, Gunter G, Hartwig EO, Valentine R (1981) Effects of chlorinated discharges on marine animals. In: Jolley RL, Brungs WA, Cotruvo JA, Cumming RB, Mattice JS, Jacobs VA (eds) Water Chlorination Environmental Impact and Health Effects, Vol. 4. Ann Arbor Science Publishers, Ann Arbor, MI, pp 947–966.

Vernot EH, MacEwan JD, Huan CC, Kinkead ER (1977) Acute toxicity and skin corrosion data for some organic and inorganic compounds and aqueous solutions. Toxicol Appl Pharmacol 42:417–423.

Vinsel PJ (1990) Treatment of acute chlorine gas inhalation with nebulized sodium bicarbonate. J Emerg Med 8(3):327–329.

Vijayan R, Bedi SJ (1989) Effect of chlorine on three fruit tree species at Ranoli near Baroda, India. Environ Pollut 57:97–102.

Warrack AJ (1978) Another domestic hazard. Med Sci Law 18(2):93–95.

Watkins CH, Hammerschlag RS (1984) The toxicity of chlorine to a common vascular plant. Water Res 8:1037–1043.

Westervelt R (1996) Chlorine production jumps. Chem Week 158(25):36.

WHO (1982) Environmental health criteria for chlorine and hydrogen chloride. World Health Organization, Geneva, Switzerland, no. 21.

Wilde EW, Soracco RJ, Mayack LA, Shealy RL, Broadwell TL (1983a) Acute toxicity of chlorine and bromine to fathead minnows and bluegills. Bull Environ Contam Toxicol 31(3):309–314.

Wilde EW, Soracco RJ, Mayack LA, Shealy RL, Broadwell TL, Steffen RF (1983b) Comparison of chlorine and chlorine dioxide toxicity to fathead minnows and bluegill. Water Res 17(10):1327–1331.

Wiley SW (1981) Effects of chlorine residuals on blue rockfish (Sebastes mystinus) In: Jolley RL, Brungs WA, Cotruvo JA, Cumming RB, Mattice JS, Jacobs VA (eds) Water Chlorination: Environmental Impact and Health Effects, Vol. 4. Ann Arbor Science Publishers, Ann Arbor, MI, pp 1019–1027.

Winternitz MC, Lambert RA, Jackson L, Smith GH (1920) The pathology of chlorine poisoning. In: Winternitz MC (ed) Collected Studies on the Pathology of War Gas Poisoning. Yale University Press, New Haven, pp 3–31. (As cited in NIOSH 1976.)

Wolf DC, Morgan KT, Gross EA, Barrow CS, Moss OR, James RA, Popp JA (1995) Two-year inhalation exposure of female and male B6C3F1 mice and F344 rats to chlorine gas induces lesions confined to the nose. Fundam Appl Toxicol 24(1):111–131.

Wood BR, Colombo JL, Benson BE (1987) Chlorine inhalation toxicity from vapors generated by swimming pool chlorinator tablets. Pediatrics 79:427–430.

Yosha SF, Cohen GM (1979) Effect of intermittent chlorination of developing zebrafish embryos (Brachydanio rerio). Bull Environ Contam Toxicol 21(4–5):703–710.

Zillich JA (1972) Toxicity of combined chlorine residuals to freshwater fish. J Water Pollut Control Fed 44:212–220.

Zimmerman PW (1955) Chemicals involved in air pollution and their effects upon vegetation. Contr Boyce Thompson Inst Pl Res 3:124.

Zwart A, Woutersen RA (1988) Acute inhalation toxicity of chlorine in rats and mice: time-concentration-mortality relationships and effects on respiration. J Hazard Mater 19:195–208.

Manuscript received March 1, 1999; accepted July 19, 2000.

# Cumulative and Comprehensive
# Subject Matter Index
# Volumes 161–170

Antarctic aerosol, anthropogenic sources, **166**:135

Antarctic climate change, trace metal changes, **166**:129 ff.

Antarctic echinoderms, metal levels, **166**:104

Antarctic ecosystems, trace element contamination, **166**:83 ff.

Antarctic fish, metal levels, **166**:106

Antarctic fish, scientific names, **166**:106

Antarctic freshwater, trace metal levels, **166**:153, 157

Antarctic, ice age estimates, **166**:87

Antarctic increasing human impact, trace metal changes, **166**:129 ff.

Antarctic inland waters, trace metal levels, **166**:157

Antarctic lakes, trace metal levels, **166**:151, 157

Antarctic marine biota, trace element levels, **166**:98

Antarctic marine mammals, trace metal levels, **166**:112

Antarctic mollusks, trace metal levels, **166**:104

Antarctic mosses/lichens, trace metal levels, **166**:118

Antarctic ozone layer changes, **166**:129

Antarctic, paleoenvironmental metal levels, **166**:85

Antarctic paradox, nutrient surplus, **166**:143

Antarctic seabirds, trace metal levels, **166**:108

Antarctic surface soils, trace metal levels, **166**:152, 154

Antarctic terrestrial biota, trace metal levels, **166**:115

Antarctic, trace element levels, **166**:85

Antarctica, anthropogenic metal contamination, **166**:158

Antarctica, atmospheric trace element deposition, **166**:132

Antarctica, gaseous mercury deposition, **166**:133

Antarctica, human activity metal contamination, **166**:158

Antarctica, ice core snow ages, **166**:130

Antarctica, map, **166**:131

Antarctica, metals recent marine sediments, **166**:148

Antarctica, trace element levels recent snow, **166**:140

Anthropogenic metal contamination, Antarctica, **166**:158

Antibiotic residues, poultry manure, **162**:135–136

Antimicrobials, categories, **164**:43

Antimicrobials, leaching potential, **164**:42

Antimony, Antarctic abiotic levels, **166**:94

Ants, arid ecosystem bioassays, **168**:64, 70

*Aphanizomenon*, toxic cyanobacteria, **163**:127

Apparent sediment-water partition coefficient, role in $K_{doc}$ determination, **169**:9

Apparent solubility, role in $K_{doc}$ determination, **169**:9

Application drift, glyphosate, **167**:47

AQUAPOL, database for soil/sediment/water partition constant, **169**:2

Aquatic acute/chronic ratios, DNA, **161**:72

Aquatic acute/chronic ratios, DNB, **161**:61

Aquatic acute/chronic ratios, HMX, **161**:113

Aquatic acute/chronic ratios, RDX, **161**:95

Aquatic acute/chronic ratios, TNB, **161**:46

Aquatic acute/chronic ratios, TNT, **161**:28

Aquatic biotoxins cyanobacteria, risks associated, **161**:182

Aquatic biotoxins finfish, risks associated, **161**:182

Aquatic biotoxins, global monitoring programs, **161**:174

Aquatic biotoxins, organisms implicated (table), **161**:163

Aquatic biotoxins, seafood safety monitoring, **161**:157 ff.

Aquatic biotoxins shellfish, risks associated, **161**:181